Transformers and Large Language Models

A Hands-On Guide to RAG and Agentic AI

Ahmed Fawzy Gad

Apress®

Transformers and Large Language Models: A Hands-On Guide to RAG and Agentic AI

Ahmed Fawzy Gad
Ottawa, ON, Canada

ISBN-13 (pbk): 979-8-8688-2784-6　　　　ISBN-13 (electronic): 979-8-8688-2785-3
https://doi.org/10.1007/979-8-8688-2785-3

Managing Director, Apress Media LLC: Welmoed Spahr
Acquisitions Editor: Celestin Suresh John
Editorial Assistant: Gryffin Winkler

Cover designed by eStudioCalamar

Distributed to the book trade worldwide by Springer Science+Business Media New York, 1 New York Plaza, New York, NY 10004. Phone 1-800-SPRINGER, fax (201) 348-4505, e-mail orders-ny@springer-sbm.com, or visit www.springeronline.com. Apress Media, LLC is a Delaware LLC and the sole member (owner) is Springer Science + Business Media Finance Inc (SSBM Finance Inc). SSBM Finance Inc is a **Delaware** corporation.

For information on translations, please e-mail booktranslations@springernature.com; for reprint, paperback, or audio rights, please e-mail bookpermissions@springernature.com.

Apress titles may be purchased in bulk for academic, corporate, or promotional use. eBook versions and licenses are also available for most titles. For more information, reference our Print and eBook Bulk Sales web page at http://www.apress.com/bulk-sales.

Any source code or other supplementary material referenced by the author in this book is available to readers on GitHub. For more detailed information, please visit https://www.apress.com/gp/services/source-code.

If disposing of this product, please recycle the paper

To Egypt, the land that shaped me, and to the people of the Arab and Islamic world. To those who recognize their true adversaries and believe that their strength lies in their unity and collaboration, not in conflict with one another or petty side disputes. May we stand as an unbroken front until our civilization reclaims its rightful place, shining as it once did, and as it must again.

Table of Contents

Table of Contents

About the Author

Ahmed Fawzy Gad, M.Sc., is a senior AI engineer and researcher based in Ontario, Canada, with experience bridging the gap between academic theory and industrial application. His career spans multiple domains in the AI landscape, including cybersecurity work at Trend Micro and applied machine learning in healthcare at Medicia Research, where he focused on oncology.

Ahmed has a strong research background with multiple peer-reviewed publications. He is the creator of PyGAD, a widely adopted open-source library for solving optimization problems using genetic algorithms.

As an educator and technical author, Ahmed has written five books covering machine learning, computer vision, and Python programming. He has also authored dozens of widely read tutorials and articles helping thousands of learners understand complex AI concepts. Known for his ability to deconstruct sophisticated architectures into clear, step-by-step explanations, Ahmed is passionate about making advanced AI accessible to both beginners and professionals, a mission he continues with this deep dive into large language models.

Ahmed is the founder of Vilvik, an early-stage Canadian startup focused on building AI-powered solutions and developing tools for optimization and intelligent systems. Currently in its initial phase, the company provides specialized AI consultation and offers an online application focused on solving optimization problems.

About the Technical Reviewer

Siddhant Agarwal is a seasoned Developer Relations professional with over a decade of experience building and scaling global developer ecosystems. He is currently a Senior Developer Advocate at ClickHouse, leading developer engagement across the APJ region, where he works at the intersection of real-time analytics, data infrastructure, and developer experience.

Previously, Sid led Developer Relations across APAC at Neo4j, where he was responsible for building and scaling the regional DevRel strategy by driving developer adoption, growing communities, launching programs and events, and representing the region across product, engineering, and go-to-market initiatives.

In his prior role, Sid has also managed flagship developer programs at Google. A recognized Google Developer Expert (GDE) in Gen-AI, he is passionate about empowering developers to reimagine how data and AI systems are built and understood.

Known for his ability to tell powerful stories through technology, Siddhant translates complex systems into narratives that inspire action and innovation. With his signature "Local to Global" approach, he helps grassroots developer communities scale their ideas into global impact. He continues to shape communities, share insights, and drive meaningful connections across the tech ecosystem.

Acknowledgments

In the Name of Allah, the Most Gracious, the Most Merciful.

Writing this book has been a journey of discovery, growth, and persistence. It would not have been possible without the grace of the Almighty and the support, guidance, and encouragement of many individuals.

All praise is due to Allah, the Lord of the Worlds. I am profoundly grateful for the strength, patience, and clarity bestowed upon me to complete this work. As it is written in the Holy Quran: *"And if you should count the favor of Allah, you could not enumerate them"* (Surah Ibrahim, 14:34), I also hold close the verse: *"My Lord, enable me to be grateful for Your favor which You have bestowed upon me and upon my parents and to do righteousness of which You approve"* (Surah Al-Naml, 27:19). Every insight within these pages is a reflection of the wisdom He provides, and any shortcoming is entirely my own.

To my mother and my family, you are my constant source of stability and my first home. Your frequent calls and check-ins were a lifeline through the most demanding hours of research and writing. Your patience and support provided the sanctuary I needed to turn this project into a reality. This book is as much yours as it is mine.

I would like to express my sincere gratitude to Frankie Wong, my former manager at Medicia Research Inc. Our time working together was formative to my professional development, and I remain deeply appreciative of the mentorship and support he continued to offer long after our formal collaboration ended. His industry insight and steady leadership have left a lasting impression and have been invaluable.

My deepest thanks go to my colleague Kawtar Harakat, a dedicated Moroccan researcher whose remarkable kindness sustained me through some of the most challenging periods of my life. Her commitment to charitable work is a constant reminder that our greatest contribution lies not only in what we know but in how generously we give of ourselves.

Beyond those named here, I extend my gratitude to the many friends, colleagues, and members of the wider community who offered encouragement along the way through a thoughtful discussion, a shared insight, or a simple word of support. No contribution was too small; each one played a part in bringing this work to life.

CHAPTER 1

Introduction to Machine Learning

Artificial intelligence is the field of embedding human-like intelligence into machines, enabling them to perform tasks efficiently that would typically require human capability. Over the years, machine intelligence has evolved from simply following human instructions to making autonomous decisions.

In this chapter, we will explore the fundamentals of machine learning by examining its typical pipeline and the types of problems it can solve. We will explore the basic model of deep learning and, for sure, the transformers model, which is the neural network.

To understand the motivation behind the transformer model, we will first review traditional recurrent neural networks and long short-term memory models, highlighting their limitations. By the end of this chapter, you will have the essential knowledge needed to begin exploring more advanced aspects of the transformer model.

1.1 What Is Artificial Intelligence?

Artificial Intelligence (AI) is a transformative field within computer science that aims to imbue machines with human-like intelligence, enabling them to perform complex tasks with little to no human intervention. At its inception, AI development heavily relied on humans to break down complex problems into structured, manageable components that machines could tackle with logical reasoning and decision-making capabilities. Over time, the machines became more intelligent to do many autonomous tasks. Throughout the evolution of AI, some subfields are introduced as illustrated in Figure 1-1.

1

© Ahmed Fawzy Gad 2026
A. F. Gad, *Transformers and Large Language Models*, https://doi.org/10.1007/979-8-8688-2785-3_1

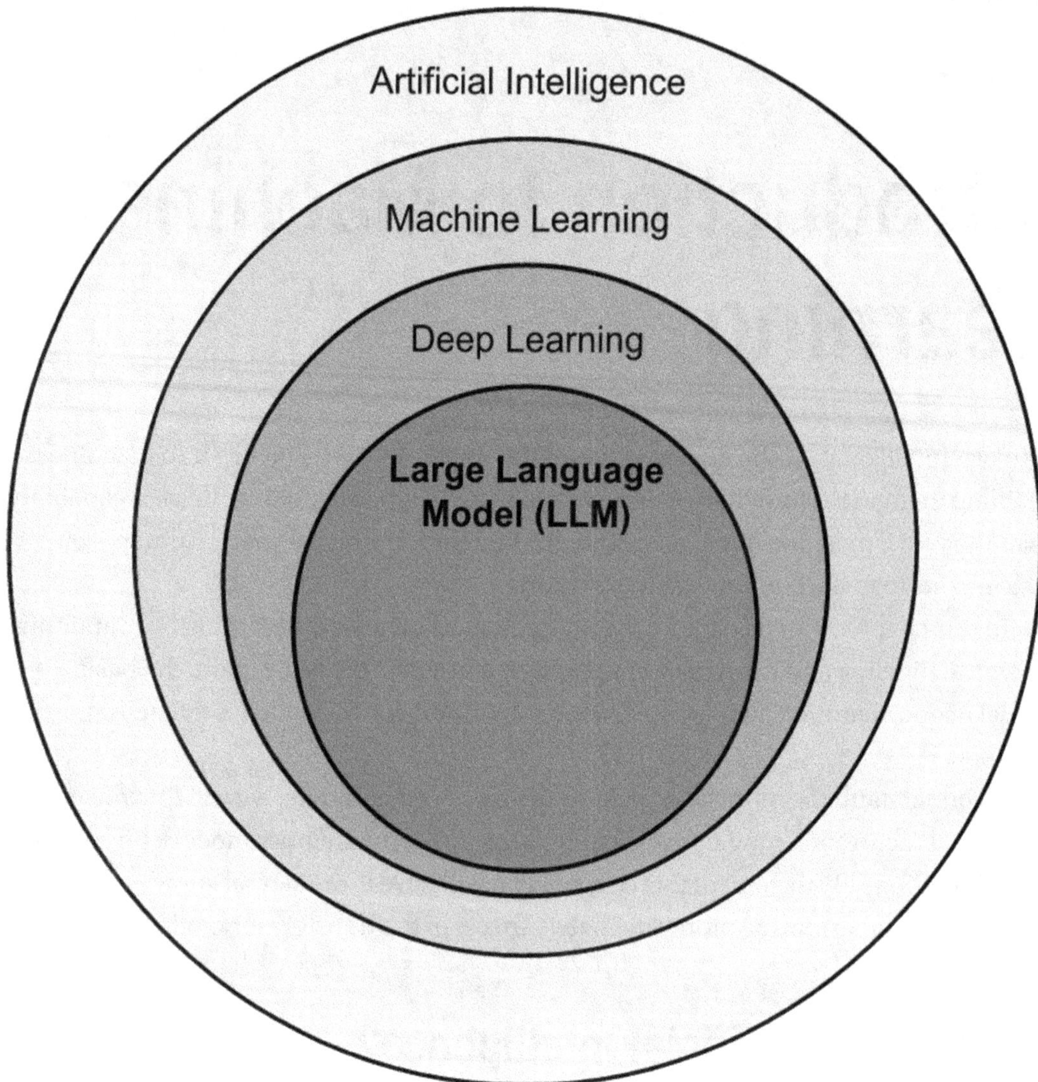

Figure 1-1. *Transformation from artificial intelligence to machine learning, deep learning, and large language models, illustrating the hierarchical relationship between these fields*

One classic illustration of AI in action is the game of tic-tac-toe (XO), played on a 3x3 grid where the goal is to align three identical symbols in a row, column, or diagonal. In this scenario, the machine is pre-equipped with human knowledge of all possible game states and potential moves. Figure 1-2 shows the possible options when the player X plays for the first time, where there is a total of 9 possible states. It also shows the states for the player O, assuming the first X is placed at the top-left corner. This makes the number of possible positions for placing the O as 8.

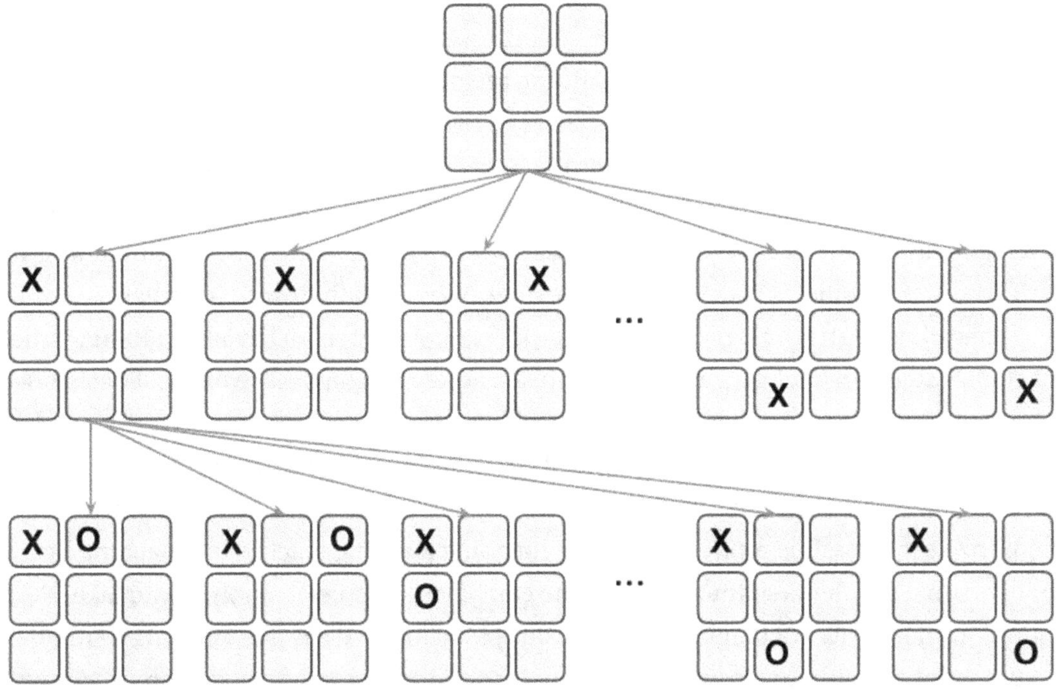

Figure 1-2. *A subset of the tic-tac-toe (XO) game state tree. A database of all possible states is generated to enable the machine to select the best move at each step*

At each turn or state, the machine evaluates its options, exploring various pathways to select the state that maximizes its chances of winning or minimizing the risk of losing. The machine looks intelligent when it plays the game and takes the correct decisions. But intelligence is hardcoded by a human, so the machine cannot go beyond it to respond to any new unseen states.

Another foundational example of early AI applications is the development of rule-based chatbots designed to assist customers with basic inquiries. These chatbots operated on predefined templates structured as sets of rules, typically using "if-else" statements. For instance, if a user greeted the chatbot with "Hi" or "Hello", the bot would respond with a welcoming message like "Hi, how can I help you?".

While these systems had limited problem-solving capabilities, they served a valuable function in routing users to human agents when queries became too complex. When provided with a description of a user's issue, the chatbot would employ basic keyword detection to identify terms such as "stopped," "failed," or "not working" to infer potential service disruptions. The bot would then collect essential information like the user's name and account number before forwarding the query to a human representative.

Simpler variations of chatbots required even less natural language processing, offering users a set of predefined options to select from. Each selection would guide the user through a decision tree, narrowing the inquiry until a final action or solution was reached. This approach ensured clarity and efficiency but underscored the limitations of early AI systems, which struggled to handle queries beyond their scripted rules.

While placing an X or O or building a basic chatbot may seem trivial to a human, the machine's ability to analyze the options, foresee outcomes, and make optimal decisions represents the core principles of artificial intelligence.

As AI continues to evolve, another sub-field called machine learning (ML) is developed that goes beyond simple rule-based systems. While AI makes the machine memorize each possible input-output pair, machine learning teaches the machine how to solve some problems so that the machine can have a degree of autonomy to take actions even if it was not taught the same example before. This is like teaching a kid how to sum numbers with examples. The teacher (or supervisor) gives the kid some examples in the form of the inputs and the expected outputs. The teacher keeps repeating such examples until they are satisfied that the student has the ability to solve the operation independently. This is done by testing the student using exams or quizzes to assess the student's ability to understand the problem.

Using machine learning, the supervisor prepares a dataset of examples for a specific problem by which the model is trained. The dataset is usually created by experts because the model's ability to solve the problem is highly affected by the quality of the data samples. There is a data engineering step that usually transforms the raw data into cleaner data that is easier for the machines to understand. After this step, each sample in the raw data is represented by a numerical feature vector that is ready to be used by the machine.

The supervisor keeps feeding the examples to the model until making sure the model understands how to solve the problem, not just memorize the examples. This is by using evaluation metrics such as the accuracy for the classification problems. Once the model is doing well on the training data, it is tested by some new unseen test dataset to make sure the model is generalized enough to handle new samples.

The field of machine learning allowed the machines to understand natural language, recognize patterns, classify problems, recognize objects in images, and more. But it is usually used to solve simple problems that are well designed by humans and have its data manually engineered to increase its quality. This almost restricted the use of machine learning to solve shallow problems. Shallow refers to the ability of the model to find the patterns that map the inputs to the outputs using few transformations.

As the volume of data generated by individuals and systems has grown, traditional machine learning approaches that rely on hand-crafted features have become increasingly impractical. Processing such vast amounts of data, often involving millions or even trillions of samples, requires models capable of learning directly from raw inputs without extensive manual feature engineering.

This challenge led to the emergence of deep learning, a subfield of machine learning that leverages artificial neural networks to automatically learn hierarchical representations from raw data. A neural network is a special machine/deep learning model that forms the base of generative AI. It will be covered later in this book.

Neural networks are still used in traditional machine learning, but such models are usually shallow with just a few layers and few neurons in each layer, as indicated in Figure 1-3.

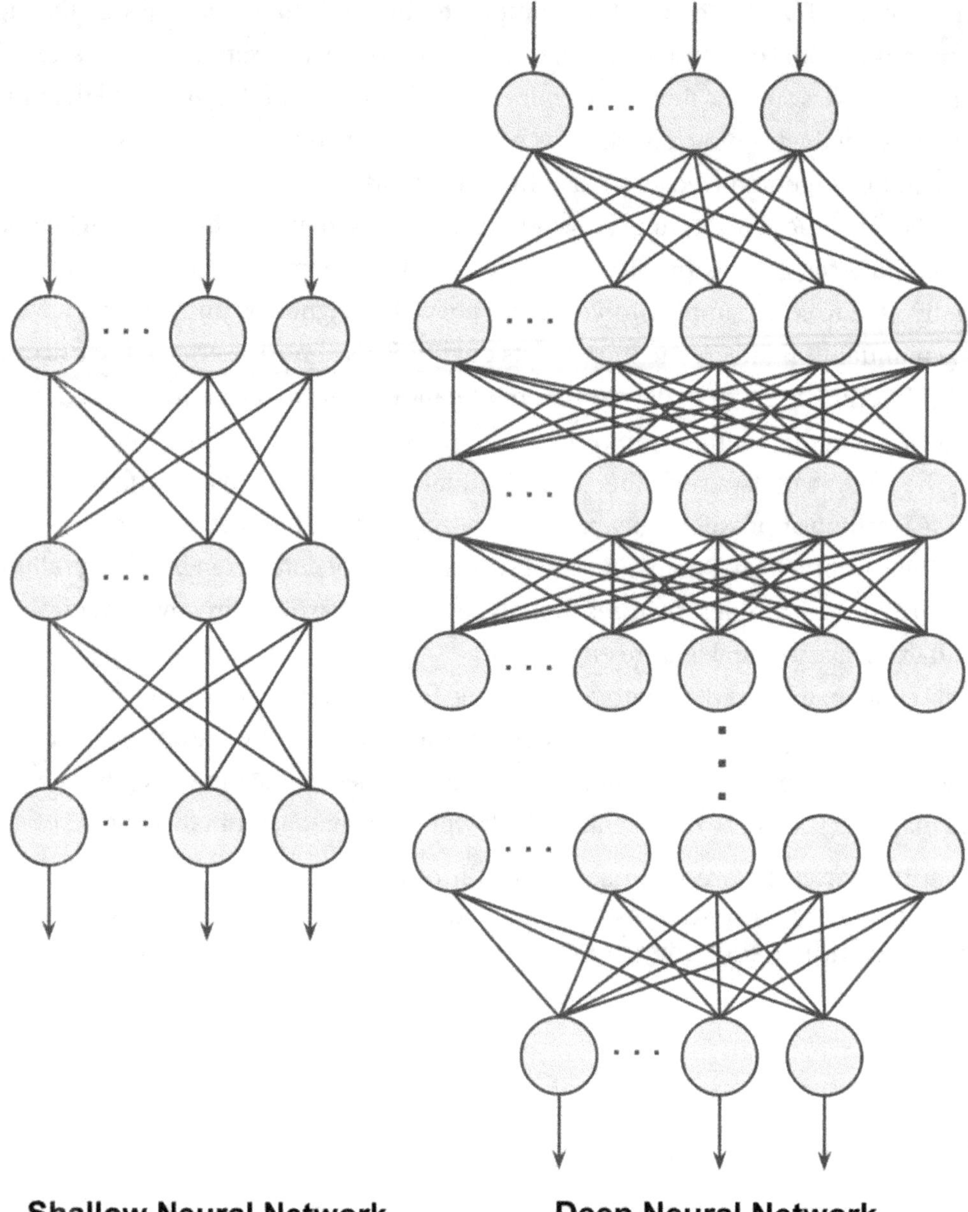

Figure 1-3. *Shallow and deep neural networks. Deep networks contain more layers than shallow networks, allowing them to learn more complex patterns*

Deep learning models are particularly effective at solving complex problems where the data require multiple layers of transformation to extract meaningful patterns and produce accurate outputs. Deep learning allowed the machines to be used in applications such as virtual assistants, autonomous vehicles, medical diagnostics, and financial forecasting.

Before the advent of the Internet and electronic devices, textbooks served as the primary source of knowledge for people. With the emergence of the Internet, text became the dominant medium of communication on websites. The rise of social media platforms further amplified the importance of text. People began sharing their experiences and opinions publicly, creating an invaluable source of real-time information. Businesses leveraged this data to gauge user satisfaction, while competitors analyzed negative feedback to improve their products and services, ultimately striving to attract and retain more customers.

As text became ubiquitous and carried vast amounts of information, it became a key objective for computer scientists to develop machines capable of understanding and learning it. The traditional natural language processing (NLP) techniques are able to understand humans' everyday languages. Through NLP, machines can perform a variety of tasks, including but not limited to:

1. **Text Classification**: Categorizing text into predefined classes.

2. **Sentiment Analysis**: Identifying and interpreting the emotional tone behind text.

3. **Text Generation**: Producing coherent and contextually relevant text.

4. **Summarization**: Condensing large volumes of text into concise summaries.

5. **Information Retrieval**: Extracting relevant information from vast data sources.

6. **Named Entity Recognition (NER)**: Identifying names, locations, dates, and other entities.

But traditional NLP models were limited to understanding only short paragraphs. As the length of the fed text increases, the models start to lose some information from the text. This heavily affected its overall understanding of the text and ability to create high-quality responses. Since users need to process large amounts of text data, the traditional small models will not work anymore. Moreover, such models are usually designed to solve a specific task. They are not general-purpose.

The previous reasons are among the drivers to use the new class of models called "large language models (LLMs)," which have been heavily used in the last few years. Such models are large because they use large amounts of data and also have large model sizes in terms of number of layers and neurons. LLMs aim to achieve a human-like comprehension of natural language, including grammar, semantics, emotions, and other linguistic aspects.

Some of these models are multimodal and able to comprehend and correlate information in different data types. An example is asking a model a textual question about an image.

Its impact is transforming the way humans interact with technology, bridging the gap between natural language and machine understanding. It is also a step toward artificial general intelligence (AGI).

1.2 Application of Generative AI

To show an example of how LLMs combine multiple applications within the same model, this is a question I asked for the Google Gemini 2.5 Flash model, accessible through this link `https://gemini.google.com`, while attaching the image in Figure 1-4.

How many times can this power bank charge an iPhone 16 Pro Max?

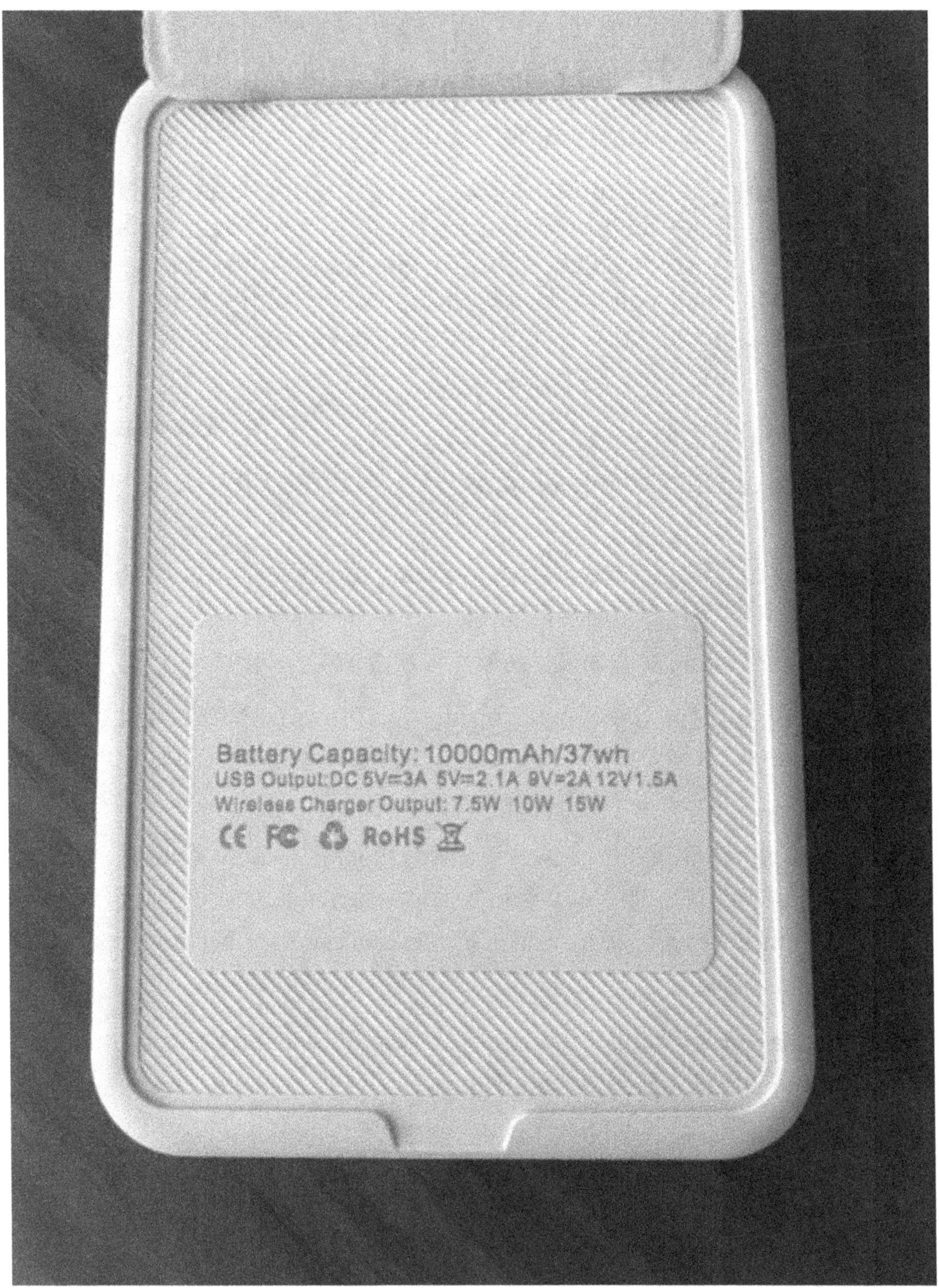

Figure 1-4. *Image of a power bank's specifications used as input for asking a question to Google Gemini, which determined that it could charge an iPhone 16 Pro Max approximately 1.8 times*

This is the model's response:

The power bank has a battery capacity of *10,000 mAh*.

The iPhone 16 Pro Max has a reported battery capacity of 4,685 mAh.

To estimate how many times the power bank can charge the iPhone 16 Pro Max, we'll divide the power bank's capacity by the phone's capacity. However, it's important to account for energy loss during the charging process (due to heat, voltage conversion, etc.). A realistic efficiency rate for power banks is typically around 80-90%. Let's use an efficiency of 85% for this calculation.

Calculation:

- Effective capacity of power bank = 10,000 mAh * 0.85 = *8,500 mAh*
- Number of charges = Effective capacity of power bank / iPhone 16 Pro Max battery capacity
- Number of charges = 8,500 mAh / 4,685 mAh $\approx$ 1.81

Therefore, this power bank can charge an iPhone 16 Pro Max approximately *1.8 times*. This means you can get one full charge and then about 80% of the second charge.

The process of generating this answer involves the integration of multiple AI applications and techniques. This multi-modal approach allows the system to bridge the gap between various data types (e.g., visual and textual) and apply complex reasoning. Key components include:

1. **Image Pre-processing**: Prior to any information extraction, an uploaded image undergoes initial processing steps. This can include quality enhancements such as noise reduction, contrast adjustment, etc. Moreover, scaling or resizing may be performed to optimize the image for subsequent computer vision models, ensuring it matches their expected input dimensions.

2. **Optical Character Recognition (OCR)**: Following image pre-processing, OCR is applied to extract textual information embedded within the image. In this specific case, OCR was crucial for accurately reading the technical specifications from the power bank's label, such as its capacity of 10,000 mAh.

3. **Named Entity Recognition (NER)**: NER is an application of NLP that helps the chatbot to identify the main entities in the prompt, which are power bank and iPhone 16 Pro Max, and the relationship between them.

4. **Web Search and Information Retrieval**: The chatbot searched the web and retrieved the capacity of the iPhone 16 Pro Max, which is 4,685 mAh. This is crucial to answering the question.

5. **Model Reasoning**: The language model thinks of how to answer the question.

6. **Generative AI**: After gathering all the necessary information, the large language model is used to generate an answer that addresses the prompt in a user-friendly format.

1.3 Machine Learning Pipeline

The ML pipeline generally consists of the following basic steps, which are applicable to NLP as well as other ML tasks:

1. Data collection

2. Data preprocessing

3. Feature engineering

4. Modeling

5. Testing and evaluation

6. Deployment

7. Postprocessing

For a machine to efficiently learn from text (or any kind of data in general), the data must be represented in a format that is machine-readable. Since machines can only interpret numerical data, text must therefore be converted into numerical representations. This is usually handled during the preprocessing step.

The process of converting textual data into numerical data involves two main steps:

1. Tokenization

2. Encoding

1.3.1 Tokenization

Tokenization is a preprocessing step responsible for breaking text into smaller units, known as tokens. In the context of NLP, a token can be defined as a word, subword, character, or symbol (e.g., end of sentence). Tokenization can vary in granularity depending on the specific task, but the goal is to create meaningful units that capture the essential elements of the text.

During tokenization, a predefined vocabulary of tokens in each language (e.g., English) is used to segment text into individual tokens. This vocabulary contains most of the common tokens in the language, mapping each to a unique token ID. The following table provides a simple example of such a vocabulary:

```
| Word       | TK-ID | Notes                             |
|------------|-------|-----------------------------------|
| hello      | 11    |                                   |
| world      | 2     |                                   |
| my         | 81    |                                   |
| name       | 4     |                                   |
| is         | 32    |                                   |
| adam       | 61    |                                   |
| how        | 17    |                                   |
| are        | 3     |                                   |
| you        | 9     |                                   |
| <unk>      | 0     | Special token for unknown words   |
```

For example, given the input sentence:

```
Hello, my name is Adam. How are you?
```

The tokenization process involves:

1. Lowercasing the text.

2. Removing punctuation (e.g., commas and periods).

3. Splitting the text into tokens based on the vocabulary.

4. Mapping tokens to their corresponding IDs.

Figure 1-5 shows the tokenized output where each token is replaced by a token ID.

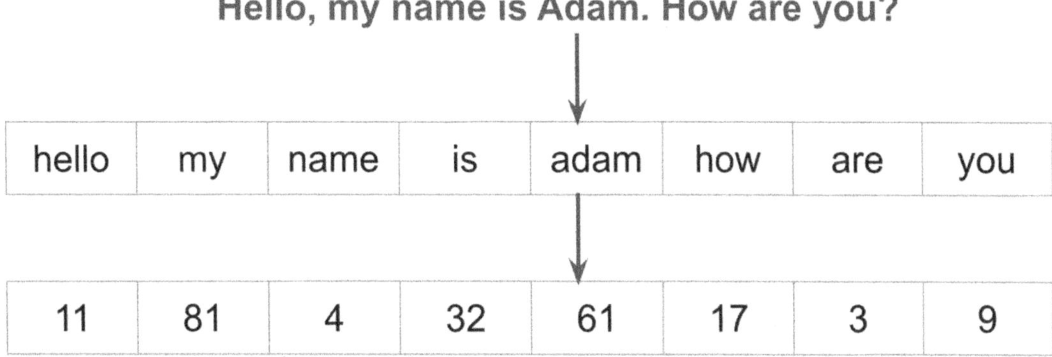

Figure 1-5. *Tokenization, the process of breaking text into smaller units called tokens, each assigned a numeric ID*

Most encoder models operate with a fixed token limit, meaning the input sequence must not exceed the model's maximum capacity. For example, GPT-3 processes up to 2048 tokens at a time. Additionally, tokenized sequences typically include special tokens to indicate the beginning and end of the input.

Beyond text, tokens can represent other data types:

- Images are tokenized into small pixel matrices.

- Audio is tokenized into short segments of sound.

Some of the text tokenization techniques are:

1. Word tokenization

2. Character tokenization

3. Subword tokenization

Word Tokenization

Rule-based tokenization applies predefined rules to split input text into separate words. A common approach involves splitting tokens based on whitespace. For example, given the sentence:

```
The teacher didn't teach yesterday.
```

It is tokenized into the following five tokens:

```
["The", "teacher", "didn't", "teach", "yesterday."]
```

However, this approach has some drawbacks. The period in *"yesterday"* is considered part of the word rather than a separate token. This issue is typically addressed through preprocessing techniques that remove punctuation marks such as commas (,) and question marks (?).

Rule-based tokenization treats *"teacher"* and *"teach"* as completely different tokens, despite sharing the same root. Since it does not account for morphological relationships, it generates a vast number of unique tokens, even for variations of the same word. As a result, word-based tokenization significantly expands the vocabulary size (reaching hundreds of thousands of tokens) where each token is encoded independently, leading to inefficiencies in text representation.

This technique also fails to account for misspelled words, as each token is processed exactly as it appears in the text. Consequently, misspellings are treated as entirely distinct tokens and encoded into separate vectors, which are then fed into the NLP model, which can lead to inefficiencies.

Character Tokenization

Character-level tokenization splits text into individual characters, treating each one as a separate token. For example, applying this method to the previous sentence produces the following token list:

```
["T", "h", "e", " ", "t", "e", "a", "c", "h", "e", "r", " ", "d", "i", "d",
"n", "'", "t", " ", "t", "e", "a", "c", "h", " ", "y", "e", "s", "t", "e",
"r", "d", "a", "y"]
```

While this approach resolves the issue of misspelled words, since individual characters are encoded rather than whole words, it introduces several challenges. Since each character must be encoded into a separate vector, the volume of training data expands significantly. Given that the average English word consists of about five characters, character-based tokenization increases the number of tokens roughly fivefold compared to word-level tokenization. This leads to a substantial rise in memory usage and training time, even for relatively small datasets.

Unlike word- or subword-based tokenization, character-level tokenization discards important linguistic structures, such as morphemes and word boundaries. As a result, the NLP model must learn words out of the characters, making training more complex and computationally demanding.

Subword Tokenization

The previously discussed tokenization techniques rely on static rules to extract tokens from text. In contrast, subword tokenization employs dynamic, data-driven rules to segment text into meaningful subwords based on statistical patterns. This approach identifies frequently occurring atomic units within a corpus, allowing for more efficient text representation.

For example, the words *capable*, *playable*, *unable*, and *avoidable* all share the subword *able*. This indicates that *able* should be treated as an independent token. Thus, a word like *playable* would be tokenized into at least two subwords: *play* and *able*.

This technique effectively handles unseen or rare words by breaking them down into familiar subword components. For example, if a user mistakenly writes *goodgame* as a single word, the algorithm will recognize that it consists of two frequently occurring subwords: *good* and *game*. This ensures that even rare or out-of-vocabulary words can be meaningfully processed, improving the model's generalization and robustness.

Once trained, the tokenizer splits text into subwords based on the learned vocabulary. Any token that is not present in the vocabulary is replaced with a special *UNK* token, ensuring that the model can handle previously unseen words.

Some of the common subword tokenization algorithms include:

1. Byte pair encoding (BPE)

2. WordPiece

3. Unigram

Byte Pair Encoding

Byte pair encoding (BPE) begins by treating each character in the dataset as an individual token. For example, in English, this would include the 26 alphabetic characters. To handle unseen characters, such as Unicode symbols and emojis, an *unknown token* (e.g., UNK) is assigned.

The algorithm then iteratively merges the most frequently co-occurring pairs of character tokens to form longer subwords. This process continues until a predefined vocabulary size is reached. BPE is used in GPT-2.

WordPiece

Developed by Google and used in BERT, WordPiece follows a similar merging process as BPE but differs in its selection criteria. Instead of using raw frequency counts, WordPiece maximizes the probability of a merged token by dividing its probability by the product of the probabilities of its individual components. This probabilistic approach enables WordPiece to construct more efficient subword representations.

$$score = \frac{P(T_1, T_2, \ldots, T_n)}{P(T_1)P(T_2)\ldots P(T_n)} \tag{1}$$

Unigram

Unlike BPE and WordPiece, which start with individual characters and iteratively merge them to form tokens, Unigram adopts a different approach. It begins with a large vocabulary of potential subwords and gradually eliminates tokens with the lowest probabilities until the desired vocabulary size is reached.

$$P(w) = \frac{\text{Count}(w)}{\sum_{\text{all words}} \text{Count}(w)} \tag{2}$$

This process ensures that the most informative and frequently occurring subwords remain.

1.3.2 Encoding

After tokenizing the text, the next critical step is feature extraction or engineering. This process involves encoding the tokens as numbers, which serve as inputs for training an ML model. These features capture essential patterns and characteristics from the data, enabling the model to learn and make accurate predictions.

This encoding (i.e., feature extraction) process involves converting each token into a vector, which is a list of numbers representing the token's properties in a multi-dimensional space.

A vector is preferred over a single number for several important reasons:

1. **Capturing Richer Properties and Features**: A vector can capture more complex features of a token. The higher the dimensionality of the vector, the more detailed the representation becomes. This allows the model to understand subtle differences in meaning or context.

2. **Creating Complex Non-linear Relationships**: Vectors enable the creation of complex, non-linear relationships between tokens. These relationships help the model understand the context in which words appear and how they interact with each other.

3. **Suitability for ML Models**: ML models, particularly deep learning models, are designed to work with vectors rather than scalars (single numbers). Vectors provide the flexibility needed for such models to learn and generalize across various tasks.

For instance, a vector representation for the word "orange" might look like the following:

```
[0.91, -0.65, 0.83, -0.21]
```

Although the individual elements of the vector are not directly interpretable, each number corresponds to a specific feature or characteristic of the word "orange." Together, these numbers form a rich, machine-readable representation of the word. The higher dimensionality of the vector enables advanced features to reflect the word's meaning in various contexts.

Some of the encoding techniques are:

1. One-hot encoding

2. Bag of words (BoW)

3. Word embedding

4. Contextual embedding

One-Hot Encoding

One-hot encoding is a fundamental technique in NLP, which begins by constructing a vocabulary containing all the unique words from the dataset. The size of this vocabulary can vary depending on the scope and diversity of the text corpus.

Once the vocabulary is established, each word (token) from the input text is represented as a binary vector with a length equal to the vocabulary size. This vector consists of zeros in all positions except for a single '1' at the index corresponding to the word position in the vocabulary. So, the only feature it extracts is whether the word exists or not.

For example, consider a vocabulary consisting of the following five words:

```
[the, play, game, test, now]
```

The word **"test"** would be represented as:

```
[0, 0, 0, 1, 0]
```

Each word is uniquely encoded in this sparse format. However, one notable limitation of one-hot encoding is that the vector size grows proportionally with the vocabulary size, which can become computationally expensive and memory-intensive when dealing with large text corpora.

Bag of Words

Unlike one-hot encoding, which produces a vector for each individual word, Bag of Words (BoW) generates a single vector representing the entire text input. The vector represents a piece of text, such as a sentence or a document, as a frequency-based vector while disregarding the order of words. Each position in this vector corresponds to a word from a predefined vocabulary, and the value at each position indicates the frequency (count) of that word in the given text. This approach not only captures the presence or absence of a word (as one-hot encoding does) but also embeds the frequency with which each word appears, providing more information about the importance of certain words in the text.

For instance, consider a vocabulary consisting of the following five words:

```
[Exciting, game, with, performance, player]
```

For the input sentence:

```
Exciting game with exciting performance
```

The resulting feature vector would be:

```
[2, 1, 1, 1, 0]
```

While the BoW and one-hot encoding approaches are simple, they can be used for tasks like comparing two documents based on word frequency. But they disregard word order, which can result in the loss of contextual meaning.

Word Embedding

Feature vectors produced using the aforementioned techniques primarily capture word presence without accounting for the relationships between words. This can lead to different sentences with distinct meanings, producing identical feature vectors. For instance, the following two sentences would yield the same representation, despite their different meanings:

```
Snow inside window
Window inside snow
```

To overcome these limitations, word embedding techniques generate dense, multi-dimensional feature vectors for each word. These vectors capture rich semantic information by representing both the word itself and its relationships with surrounding words. Words with similar meanings yield similar vectors, enhancing a model's ability to identify word similarity and improve natural language understanding.

A widely used traditional word embedding method is Word2Vec, which offers two primary architectures for creating word vectors: continuous bag of words (CBoW) and Skip-Gram. Each of these architectures employs a shallow neural network with a hidden layer that functions as an embedding layer to learn word representations.

> Neural networks form the foundation of deep learning. Understanding how it works is crucial to learn more advanced models, such as transformers.

The CBoW architecture aims to predict a target word based on its surrounding context words. The number of context words is determined by a fixed window size. For instance, if the window size is 2, the model will use two words to the left and two words to the right of the target word for prediction.

Consider the following sentence fragment:

```
Quick brown fox jumps over ...
```

To train the model to predict the word fox, the two words to the left [quick, brown] and the two to the right [jumps, over] serve as inputs to the neural network. Each word is represented as a one-hot vector, with a length equal to the vocabulary size. If the vocabulary consists of five words, each vector will have a length of five.

Each of the four one-hot vectors is used to index an embedding matrix containing the initial word embeddings. This process retrieves the dense embedding vectors corresponding to the input words. The element-wise average of these four vectors is computed to form a single input vector for the neural network. During training, the embedding vectors are adjusted, and the model gradually improves its predictions, leading to more accurate and semantically meaningful word embeddings.

Contextual Embedding

Although word embedding techniques generate vectors that can capture contextual relationships (albeit with certain limitations), these vectors remain static once the neural network has been trained. For example, the embedding for the word *apple* will always be the same, regardless of its contextual use in the 2 statements as in Figure 1-6:

1. The apple is delicious.

2. The apple device is powerful.

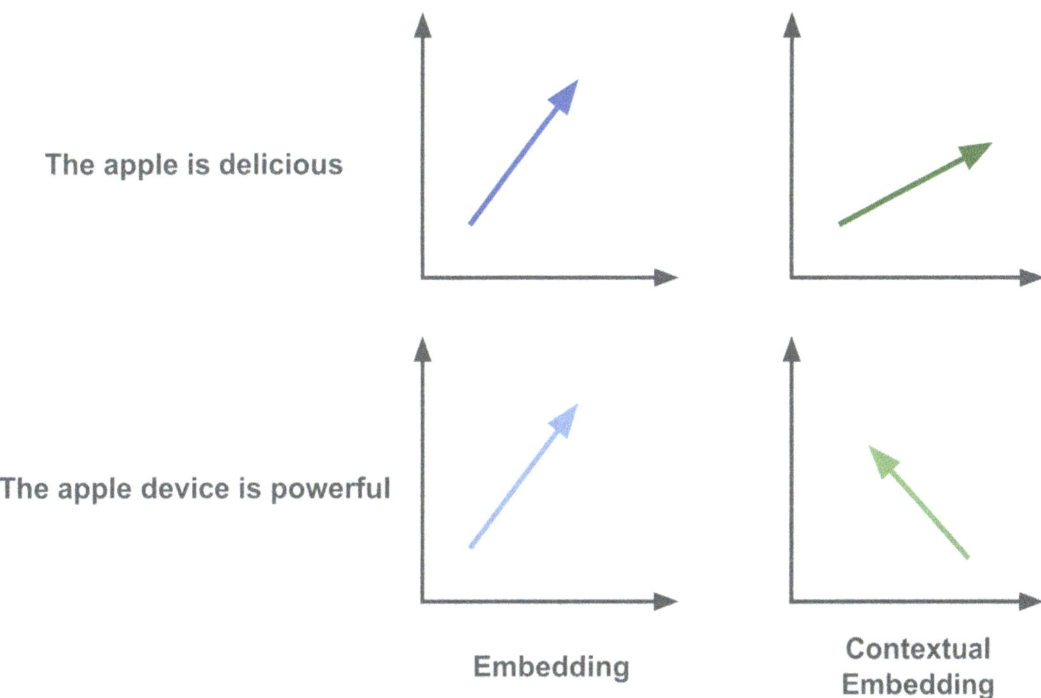

Figure 1-6. *Comparison of embedding and contextual embedding, showing that contextual embedding captures how the same word apple can have different meanings depending on context. For simplicity, the embedding vector is shown with a length of 2*

In the training data, the word *apple* may have primarily been associated with other fruits. As a result, the embedding vector for *apple* would not adapt to a technological context. This limitation impedes the model's ability to predict that the missing word could be *apple* in a context related to technology.

Contextual embeddings address this issue by considering the entire context when generating the embedding vector for a token or word. Unlike traditional embeddings, which are static, contextual embeddings are dynamic and change based on the surrounding words, as in Figure 1-6, where the vector of the word *apple* points to different directions according to the context. This approach is central to state-of-the-art transformer models, which are built on the attention mechanism. These models are capable of understanding a word's meaning by analyzing its full context, rather than relying on a fixed representation.

An example of a model that uses transformers and contextual embeddings is BERT (Bidirectional Encoder Representations from Transformers). BERT's bidirectional architecture enables it to capture the meaning of words by considering both the preceding and succeeding words in a sentence. The attention mechanism, which will be discussed in detail later in this book, plays a key role in this process.

As the quality of the extracted feature vector (i.e., the embedding vector) improves, the performance of various NLP applications, such as classification, becomes more accurate and semantically meaningful. Higher-quality embeddings capture richer, more nuanced representations of words, leading to better understanding and interpretation in downstream tasks.

1.4 Modeling

Once the text data has been transformed into numerical feature vectors, the next step in the ML pipeline is modeling. This stage involves selecting or designing an appropriate model and training it using the vectorized data.

Machine learning models are generally divided into two primary categories:

1. **Supervised Learning**: The model is trained on labeled data where each input is paired with the correct output. The goal is to learn mapping from inputs to outputs.

2. **Unsupervised Learning**: The data has no labels, and the model attempts to discover hidden patterns or structures within the input data, such as clusters.

A learning model is considered unsupervised when it receives no guidance or supervision during the training process. One common form of supervision is the assignment of labels to each data sample, which helps the model explicitly understand the class or category each sample belongs to. If the dataset lacks such labels, the learning process is classified as unsupervised learning. On the other hand, when the data includes labels that specify the correct output, the process is known as supervised learning.

While there are additional types of ML, such as reinforcement learning, where an agent learns by interacting with an environment, we will focus primarily on these two core categories.

1.4.1 Supervised Learning

Supervised learning is typically used for problems where the data samples share a consistent structure and follow general properties that remain relatively stable over time. To improve model performance in these cases, it is often worthwhile to invest time and effort into creating a labeled dataset. This allows the model to learn how to distinguish between different types of samples based on their associated labels.

For example, humans and vehicles have distinguishing characteristics that are generally consistent. When building a computer vision model to detect humans and vehicles in images, the most effective approach is to use a labeled dataset that includes examples of both classes. Each sample in the dataset is labeled to indicate whether it represents a human or a vehicle. The supervised learning model then uses this labeled data to learn how to differentiate between the two classes.

Supervised machine learning models can be categorized in various ways, but one of the most important criteria is the type of output the model is designed to predict, as in Figure 1-7:

1. **Classification**: The model predicts a discrete label chosen from a finite set of possible categories.

2. **Regression**: The model predicts a continuous numerical value.

In a classification problem, the total number of possible outcomes is known and limited. When the number of possible outputs is effectively unlimited or spans a continuous range, the task is considered a regression problem.

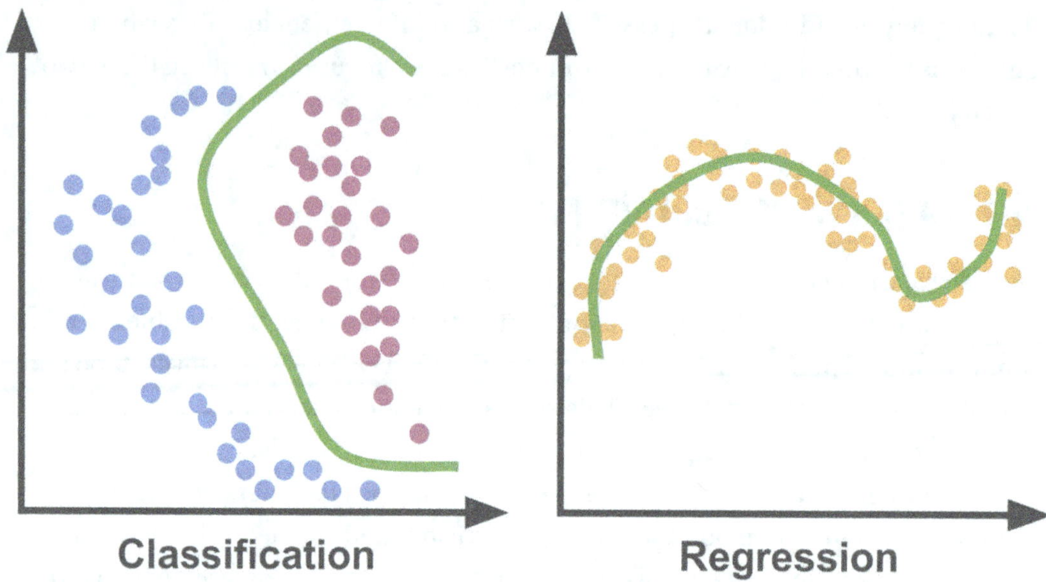

Figure 1-7. *Classification is about achieving the best decision boundary that separates the samples of the different classes or groups to maximize accuracy. Regression is finding the fittest line or curve that captures the relationship between the input features and the continuous output to minimize the error*

Several widely used models fall under the umbrella of supervised learning. Common examples include:

1.   Artificial neural networks

2.   Random forests

3.   Support vector machines

4.   Decision trees

Artificial neural networks (ANNs) are computational models inspired by the biological neural networks in human brains, consisting of layers of interconnected cells called neurons.

Each connection has an associated weight that is iteratively adjusted during training to minimize a loss function such as cross-entropy for classification or mean squared error for regression.

Because they can stack many hidden layers, neural networks are uniquely capable of modeling highly non-linear and intricate patterns. This flexibility allows them to excel at everything from predicting house prices to identifying objects in images or processing natural language.

Support vector machines (SVMs) work by mapping data points into a high-dimensional space to find the optimal hyperplane that separates or fits the data with the maximum margin.

For classification (support vector classification [SVC]), the goal is to find a boundary that maximizes the distance between different classes.

For regression (support vector regression [SVR]), the goal is to find a hyperplane that contains as many data points as possible within a specific threshold. Through the use of kernels, SVMs can efficiently handle non-linear relationships by implicitly transforming the data into even higher dimensions.

Random forests (RFs) are an ensemble learning method that operates by constructing a multitude of decision trees at training time. To make a prediction, the model aggregates the results from all individual trees using a majority vote for classification or an average of the results for regression.

The bagging technique (bootstrap aggregating) significantly improves predictive accuracy and controls overfitting by ensuring that the model doesn't rely too heavily on any single feature or data point.

A decision tree (DT) is a non-parametric supervised learning model that predicts the value of a target variable by learning simple decision rules inferred from the data features.

DT works by structuring some rules into a tree-like flowchart where each node represents a test on an attribute, and each branch represents the outcome of that test.

In a classification context, the tree leaves represent class labels, whereas in regression, they represent continuous numerical values such as the mean of the observations in that leaf. While highly intuitive and easy to visualize, single trees are sensitive to small changes in data and can easily overfit.

These models can be applied to both classification and regression tasks, depending on how they are configured and trained.

Neural networks form the foundation of modern deep learning architectures, including transformers. A solid understanding of how neural networks work is essential for exploring more advanced models.

1.4.2 **Unsupervised Learning**

Unsupervised learning is often used when it is difficult or impractical to define a consistent set of properties that apply to each sample in the target data. One common reason for this challenge is that data may evolve over time. Features or patterns that are meaningful in the current dataset may no longer be valid after a certain period. This could be weeks, months, or years. In such cases, manually labeling the data or designing fixed rules becomes unreliable, and unsupervised learning offers a flexible alternative by discovering patterns directly from the data without the need for labeled examples.

A common example is building a spam detection system to classify emails as either spam or legitimate. Attackers frequently change their writing style, vocabulary, and patterns to evade detection. If the spam detector is trained solely on older data, it may fail to recognize newer forms of spam that differ from previously seen examples. To remain effective, the model must be continuously updated with new labeled samples that reflect the evolving nature of spam. However, even with regular updates, attackers can persistently adapt their techniques, engaging in a continuous game of evasion.

In such dynamic environments, unsupervised learning can offer a robust alternative. One effective unsupervised technique is data clustering, where the model groups unlabeled data into clusters based on similarity, as in Figure 1-8. Each new email is assigned to a cluster by comparing it to existing samples, without relying on predefined labels. Clustering is especially useful when data characteristics change over time, making it difficult to maintain an accurate, labeled dataset. This approach allows the system to detect new or unusual patterns, potentially flagging novel forms of spam.

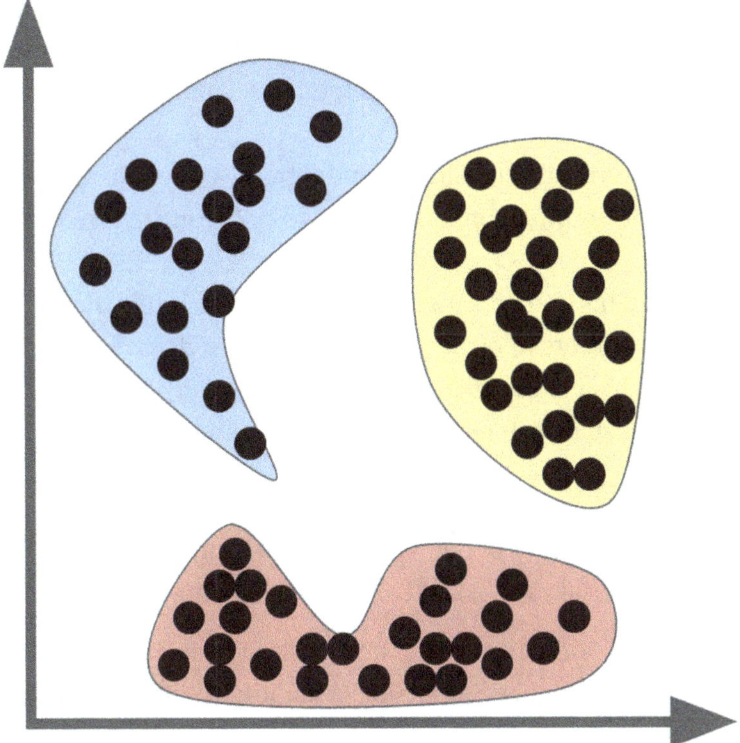

Figure 1-8. *Data clustering identifies groups (or clusters) of similar samples, where each sample is assigned to the cluster containing the samples most similar to it*

Some of the data clustering models are:

1. K-Means

2. Hierarchical

3. DBSCAN (density-based spatial clustering of applications with noise)

K-Means is a centroid-based algorithm that partitions data into K distinct, non-overlapping clusters. It works iteratively by assigning each data point to the nearest cluster center (mean) and then updating the centers based on the average of the assigned points.

While highly efficient for large datasets, K-Means assumes that clusters are spherical and of similar size, which can lead to poor results if the data has complex shapes or varying densities.

Hierarchical clustering builds a multi-level hierarchy of clusters rather than a single partition. It is typically implemented as agglomerative (bottom-up), where each data point starts in its own cluster and pairs are merged based on similarity until one large cluster remains.

The results are often visualized using a dendrogram, a tree-like diagram that allows a researcher to decide the optimal number of clusters by cutting the tree at a specific level of granularity.

DBSCAN groups points that are closely packed together while marking points in low-density regions as outliers or noise. Unlike K-Means, it does not require the user to specify the number of clusters in advance and is exceptionally good at finding clusters of arbitrary, non-spherical shapes. It relies on two key parameters:

1.   The search radius.

2.   The minimum number of points required to form a dense region.1.4.3 Causal vs. Masked Language Modeling

Having explored traditional supervised and unsupervised learning models, we now turn to transformer-based language modeling, a paradigm shift that has redefined how machines process human language.

At its core, the transformer architecture relies on a self-attention mechanism to analyze entire sequences of text simultaneously, allowing it to understand the relationship between words regardless of how far apart they are in a sentence.

While classical supervised learning depends on human-annotated labels and traditional unsupervised learning focuses on identifying clusters or patterns in data, transformer-based modeling typically utilizes self-supervised learning. In this approach, the model generates its own supervision by hiding parts of the input data and attempting to predict them.

By generating more samples out of the single training sample, the transformer model is effectively being trained on massive unlabeled datasets to learn the underlying structure of language.

There are two fundamental types of transformer-based language modeling, as illustrated in Figure 1-9:

1.   Causal

2.   Masked

Causal language modeling is used for text generation, where the model attends only to previous tokens (i.e., tokens to the left) to predict the next token in the sequence. This is the classic approach to language modeling and is employed in GPT-based models.

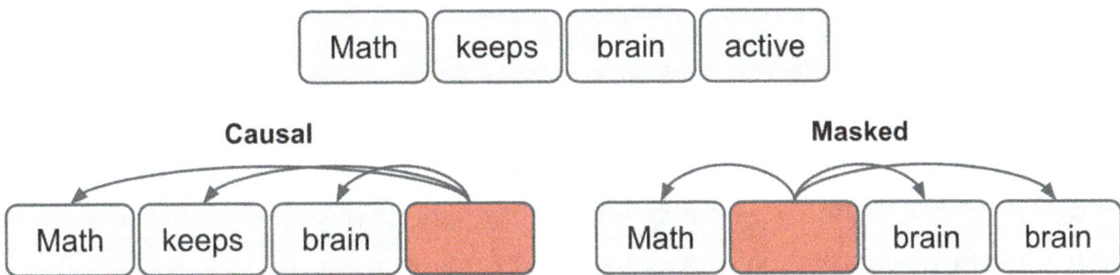

Figure 1-9. *Comparison between causal and masked language modeling*

Masked language modeling is used for predicting missing tokens within a sequence. In this approach, the target token is masked, and the model attends to tokens on both sides (i.e., bidirectionally) to predict the masked token. This strategy is used in BERT-based models.

From these two primary approaches, other variations arise, such as:

- **Sequence-to-Sequence Modeling**: Mapping an input sequence to an output sequence (e.g., machine translation).

- **Sequence Classification**: Assigning a label to an entire input sequence (e.g., sentiment analysis).

1.5 Feed-Forward Neural Network

A feed-forward neural network (FFNN) is a type of machine learning model inspired by the function of biological neural networks in the human brain. Its strength lies in its ability to process large volumes of data, making it a foundational model in deep learning (DL). DL extends traditional ML by building deep neural network models, which consist of hundreds or even thousands of layers, enabling them to learn complex patterns and representations from vast datasets. A good understanding of how FFNNs work is essential to dive deeply into advanced models like transformers.

A feed-forward neural network is just one type of neural network. Other types include recurrent neural networks, graph neural networks, convolutional neural networks, etc.

As in Figure 1-10, the FFNN consists of three types of layers:

1. Input

2. Hidden

3. Output

It is common practice to represent the neurons in the graph as a circle. To represent data flow from one neuron to another, it is represented as a line connecting the two neurons. Each connection represents a trainable parameter (i.e., weight) that, after being trained, stores the information learned by the network. The objective is to train such weights to minimize error or loss in predicting the desired outputs.

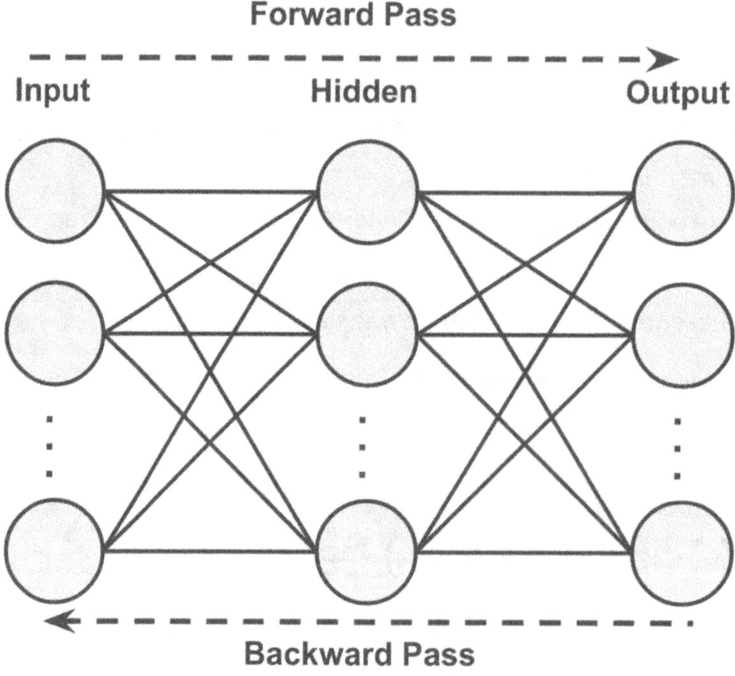

Figure 1-10. *Feed-forward neural network, where information flows in one direction from the input layer through hidden layers to the output layer without cycles or loops*

The input layer receives data in the form of vectors. In the context of NLP, embedding vectors can serve as the input to the network. These inputs are then passed through one or more hidden layers for processing. The hidden layers act as transformation functions, mapping the inputs from their original space into a new space where patterns and relationships in the data are easier to extract and learn.

The output from the final hidden layer is passed to the output layer, which produces the network's final prediction. The output layer can return a single value, multiple values, integers, or decimals, depending on the nature of the application.

For instance, in classification problems, the number of neurons in the output layer typically corresponds to the number of classes, where each neuron represents a class. The outputs of the neurons are often represented as probabilities across classes when using activation functions like softmax, which is common in multi-class classification tasks. The predicted class is the one corresponding to the neuron with the highest probability.

The network is trained through multiple iterations, where each iteration consists of two passes:

1. **Forward Pass**: The input data is propagated through the network.

2. **Backward Pass**: The loss is propagated back through the network.

1.5.1 Forward Pass

During the forward pass, the input data is propagated through the network layer by layer. Each hidden neuron computes the weighted sum of its inputs and the corresponding connection weights, as described in Equation 1. This weighted sum is then passed through an activation function such as ReLU (rectified linear unit) or Sigmoid. The activation function enables the network to model complex patterns and capture non-linear relationships in the data, which is crucial for solving real-world problems.

$$s = \sum_{i=1}^{n} w_i x_i + b \tag{3}$$

Equation 1 Sum of products between the inputs and the weights, where w_i is the weight, x_i is the input, b is the bias, and n is the number of neurons.

Each hidden and output neuron is typically assigned a bias term, which acts as an additional input with a constant value (often not shown in network diagrams). The term b in the equation represents the bias weight. It is associated with a constant input, typically set to 1. The bias allows the neuron to shift the decision boundary along the horizontal axis, providing greater flexibility in learning the patterns in the data.

The forward pass ends by calculating the network loss. An example of a commonly used loss function is the root mean square error (RMSE), which measures the average magnitude of the errors between predicted and actual values.

Assuming there is a network with 10 inputs, a single hidden layer with 5 neurons, and 2 outputs. The total number of trainable parameters or weights is $10 * 5 + 5 + 5 * 2 + 2 = 67$.

1.5.2 Backward Pass

Once the loss is calculated, the backward pass (backpropagation) begins to adjust the network's weights to minimize the loss. Gradient-based optimizers are usually used during neural network training. Popular gradient-based optimization algorithms include stochastic gradient descent (SGD), Adam, and root mean squared propagation (RMSprop), each offering different strategies to improve convergence speed and stability.

The training process relies on the chain rule as in Equation 2 to compute the partial derivatives of the loss with respect to each weight, tracing backward from the output layer to the input layer. The goal is to determine the gradient of the loss with respect to the weights, which indicates the direction and magnitude of change needed in each weight to reduce the loss.

$$\frac{\partial L}{\partial w} = \frac{\partial L}{\partial y} \frac{\partial y}{\partial z} \cdots \frac{\partial s}{\partial w} \tag{4}$$

Equation 2 The chain rule to calculate the gradient of the loss L w.r.t. a single weight w, where s is the weighted sum between the inputs and the weights, y is the output of the activation function of a neuron, and z is the weighted sum of that neuron (i.e., input of the activation function).

By iteratively performing forward and backward passes (often hundreds or thousands of times), the network gradually updates its weights toward values that minimize the loss. Ultimately, this results in a model with optimized weights capable of making accurate predictions.

1.5.3 Limitations

FFNNs are powerful tools for learning data structures, especially when a deep network is used to capture subtle features and complex relationships between inputs and outputs. However, FFNNs have several limitations that make them less suitable for tasks involving sequences:

1. No memory

2. No loops

3. Inability to handle variable-length inputs or outputs

4. The input is processed at once

Traditional FFNNs lack the ability to store learned information or capture dependencies in the data because they do not have an inherent memory mechanism. Each input sample is treated independently, without considering the relationships between samples. This limitation makes FFNNs less suitable for tasks where understanding temporal or sequential dependencies is crucial, such as in time-series or NLP tasks.

FFNNs are called feed-forward because the data flows in one direction from the input layer to the output layer, passing through one or more hidden layers. Once the data is processed by a layer, it does not revisit that layer, as the network does not have any loops or cycles. The backward pass does not involve the flow of data through the network. Instead, it propagates gradients to adjust the weights. Even if they have a memory, FFNNs lack the mechanism to cycle the learned information inside the memory back through the network.

FFNNs are designed to work with predefined inputs and outputs, where the input length must match the number of input neurons, and the output length is fixed according to the number of neurons in the output layer. The input must be available in advance before being fed into the network. This makes FFNNs well-suited for tasks with static-sized data, such as tabular data, where each record has a fixed number of features, and the task involves producing a fixed number of outputs.

The static structure of FFNNs is not well-suited for tasks where the input or output sequences have variable lengths, as in time-series or NLP tasks, or if the input data is partially available or arrives sequentially over time. For example, different text samples can have varying numbers of tokens, but the network must still handle these variable-length inputs. To address this, extended versions of FFNNs are used, including:

1. Recurrent neural networks (RNN)

2. Long short-term memory (LSTM)

3. Gated recurrent unit (GRU)

Understanding the basic principles of RNNs is highly recommended, as this provides a foundation for comprehending more advanced models, such as the transformer model with its attention mechanism.

1.6 Recurrent Neural Network

Recurrent neural networks (RNNs) differ from traditional FFNNs in their ability to handle sequential data. While FFNNs accept fixed-size inputs and produce fixed-size outputs, RNNs can process sequences of variable lengths as both inputs and outputs. This makes RNNs particularly suitable for tasks involving sequences, such as translating a sentence from English to French or analyzing a series of time-dependent data points.

Here are some of the key differences between RNNs and FFNNs:

1. **Sequential Nature**:

   - FFNNs treat each input sample as independent of the others.

   - RNNs model dependencies between inputs where each input in a sequence is influenced by the preceding inputs.

2. **Order Sensitivity**:

   - In FFNNs, the order of inputs does not matter because each input is processed in isolation.

   - In RNNs, the order of inputs is crucial, as it affects the relationships the network learns across the sequence.

3. **Loops and Memory**:

- RNNs are called recurrent because they recycle information over time through loops. The output of an RNN cell (hidden state) can be fed back as input to the next cell, enabling the network to retain memory of previous steps.

- FFNNs, in contrast, process data in a strictly forward manner, from input to output, without feedback loops or any form of memory.

This ability to carry information across time steps allows RNNs to solve sequential problems, which FFNNs are inherently incapable of handling due to their static and memoryless structure.

Figure 1-11 highlights the key difference between FFNNs and RNNs, where the RNNs have a feedback loop allowing them to process sequential inputs with dependencies. In FFNNs, the output y of a hidden neuron is calculated by applying an activation function over the sum of products of the inputs and weights. Once the output is passed to the next layer, it is discarded and never used again.

In contrast, in RNNs, this output is referred to as the hidden state y. It is saved into memory to use it later. If there are five hidden neurons, then their five hidden states are saved into the memory.

To compute the actual output y of a hidden neuron in an RNN at time step t, the hidden state h from the previous time step $t - 1$ is multiplied by a new weight and then fed (along with a bias term) into an activation function.

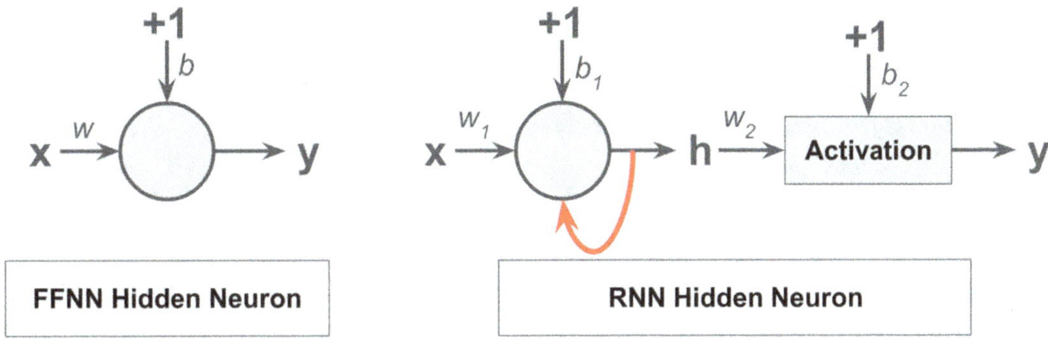

Figure 1-11. *Comparison between a hidden neuron in a feed-forward neural network and a recurrent neural network, highlighting how RNN neurons maintain connections to previous states to capture sequential dependencies*

At the beginning of the sequence, there is no previously generated hidden state. To address this, the initial hidden state is typically initialized to zero. This ensures that the first hidden state is computed solely based on the input, without any influence from previous time steps, thereby avoiding unintended biases in the learning process.

Figure 1-12 illustrates the architecture of an RNN as compared to the FFNN. Each hidden neuron produces a hidden state h_t at time step t, which is fed back as an input to the same hidden neuron at the next time step $t + 1$. The output at each time step is obtained by applying an activation function to the hidden state h_t, after it has been multiplied by an output weight.

If the RNN consists of multiple hidden layers, the hidden state from layer l serves as the input to layer $l + 1$ at each time step. This hierarchical structure allows the network to capture increasingly complex patterns over time.

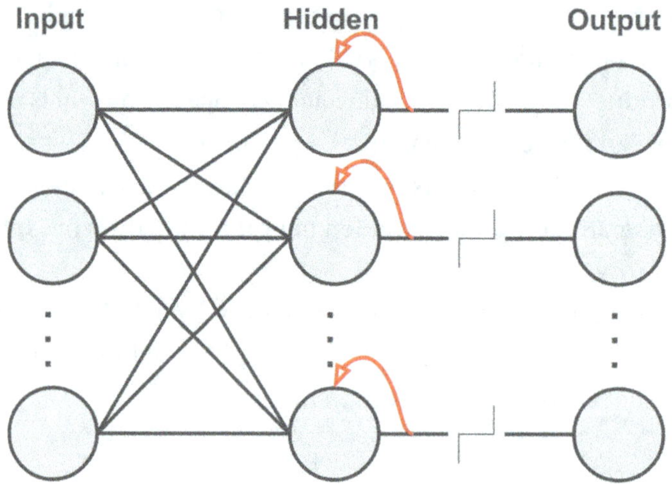

Figure 1-12. *Recurrent neural network with feedback loops where information from previous time steps influences the current output*

It is a common practice to represent the RNN where each time step t is illustrated as a separate RNN cell as shown in Figure 1-13. At each time step, an RNN cell generates a hidden state vector h_t, representing the output from all hidden neurons at that step (as depicted in Figure 1-12).

This hidden state vector h_t is passed to the next time step alongside the next input in the sequence x_{t+1}. This process repeats until all inputs in the sequence are processed. It is important to note that although the figure depicts the RNN cells as separate units across time steps, these cells are not independent. In reality, the same RNN cell is reused at each time step, meaning that the parameters (i.e., weights and biases) are shared across all time steps. This parameter sharing is a fundamental characteristic of RNNs, enabling them to generalize across sequences of different lengths while keeping the number of trainable parameters fixed.

Each hidden state h_t represents the context accumulated from all previous inputs in the sequence up to the time step t, not just the latest input x_t. When this hidden state is passed to the next RNN cell, it allows the network to retain information from earlier time steps, capturing the dependencies across the sequence.

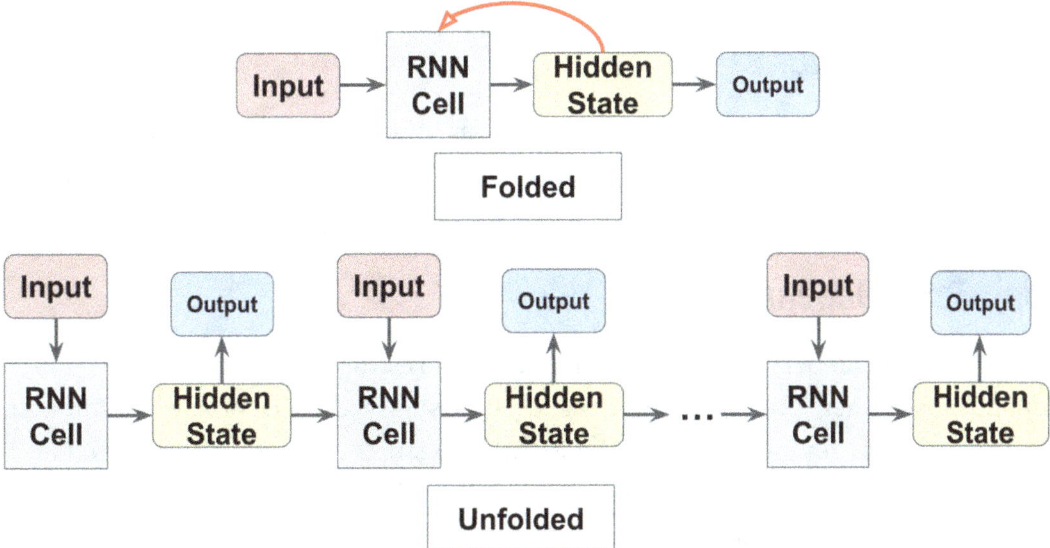

Figure 1-13. *Recurrent cell in its folded form and its unfolded representation showing how the same cell processes sequential inputs*

At each time step t, the RNN generates a hidden state vector h_t from the last hidden layer, where the length of the vector is equal to the number of hidden neurons in that layer. The number of hidden state vectors produced corresponds to the number of inputs in the input sequence. For example, if the RNN processes an input sequence of 10 tokens, and the last hidden layer contains 15 neurons, the RNN will generate a sequence of 10 hidden state vectors, each of length 15.

The way these hidden states are used depends on the nature of the task being addressed:

- For classification tasks, the hidden state vector at the final time step h_t is typically fed into a dense (fully connected) layer, followed by an activation function (e.g., Softmax) to produce class probabilities.

- For text generation or sequence prediction tasks, the hidden state vector at each time step h_t can be fed into a dense layer to generate a probability distribution over the next word in the sequence.

1.6.1 Drawbacks

Even though RNNs are flexible to handle a wide variety of sequence-based tasks, including many NLP and time-series problems, they have some drawbacks that limit their usability at a certain level.

Long-Term Dependencies

While RNNs can model short-term dependencies, they struggle with long-term dependencies, where information from earlier time steps fades as it propagates through the sequence. The reason is that all previous information is compressed and stored in a hidden state. As new information is continuously fed into the network, the hidden state gradually forgets the older information and shifts its focus toward the more recent inputs.

To clarify this limitation, consider an example where an RNN processes a sequence of 10 sentences. Suppose the first sentence introduces a piece of important information, such as the name of a person or object. Later, in the final sentence, the RNN is required to recall and fill in that name based on its memory. Since the RNN primarily retains short-term memory, it may struggle to produce the correct name. This is because the information from the later sentences has effectively overwritten or diluted the earlier information stored in the hidden state.

Vanishing and Exploding Gradients

RNNs also suffer from vanishing and exploding gradient problems. These issues arise when the gradients calculated during the backward pass become either extremely small, causing the network to nearly stop learning, or excessively large, leading to large steps in the loss space and making learning unstable.

Assuming the following values, with all weights greater than 1:

- Hidden state value: $h_i = 1$

- Hidden state weight: $W_{hi} = 2$

- Current sequence input value: $d_i = 0.5$

- Input weight: $W_{di} = 4$

- Bias weight: $W_b = 1$

The sum of products (SOP) at this time step is calculated as:

$$SOP = h_i W_{hi} + d_i W_{di} + W_b \tag{5}$$

Substituting the given values:

$$SOP = 1 \cdot 2 + 0.5 \cdot 4 + 1 = 5 \tag{6}$$

Assuming the activation function is ReLU, the next hidden state value becomes $h_{i+1} = 5$.

This updated hidden state, along with the next input from the sequence, is used to calculate the hidden state at the next time step. Suppose the new input is $d_{i+1} = 2$, the SOP is calculated as:

$$SOP = 5 \cdot 2 + 2 \cdot 4 + 1 = 19 \tag{7}$$

The next hidden state becomes $h_{i+2} = 19$.

Since all the weights are greater than 1, the hidden states increase rapidly with each time step. Over long sequences, the hidden states can grow to extremely large values.

During the backward pass, gradients are computed using the chain rule. The gradient at each weight depends on the hidden states from previous time steps. When hidden states are large, the individual partial derivatives become large. Multiplying these large partial derivatives during backpropagation results in an excessively large gradient. As the network undergoes repeated multiplications over time, the gradient increasingly grows.

Such a large gradient can cause drastic updates to the network's weights. For example, if a weight was previously 0.5, it might become 10 after an update. This causes the network to make large jumps, causing unstable training. This problem is known as exploding gradients.

Conversely, if the weights are all smaller than 1, the SOPs, and hence the hidden states, decrease over time. This leads to small partial derivatives and ultimately a small gradient. When this small gradient is applied during weight updates, the changes to the weights are minimal. As a result, the network takes tiny steps in the loss space, causing very slow learning or even stagnation. This is known as vanishing gradients.

These issues are particularly problematic in RNNs when processing long sequences, motivating the development of advanced architectures like long short-term memory (LSTM) and gated recurrent unit (GRU).

1.7 Long Short-Term Memory

The long short-term memory (LSTM) model is an improvement of the RNN to fix three main issues:

1. Vanishing gradient

2. Long-term dependencies

3. Memory retention

In addition to the short-term memory (i.e., hidden state) used in RNN, LSTM introduces long-term memory (i.e., cell state) to store long-term information. It is a vector of length equal to the number of previous units. This is the primary contribution of LSTM. The architecture of the LSTM cell is shown in Figure 1-14.

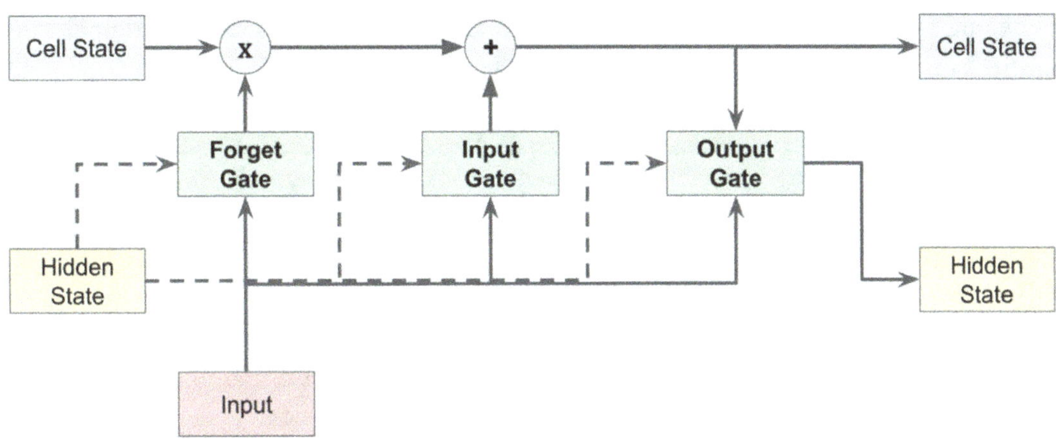

Figure 1-14. *Long short-term memory (LSTM) cell illustrating its internal gates and memory cells*

LSTM uses three gates to control the amount of information to store or discard in the forward pass:

1.   Input

2.   Forget

3.   Output

Each gate accepts the hidden state and the current input from the sequence as inputs. The output gate also receives the cell state as an input to produce the new hidden state.

1.7.1 Forget Gate

The forget gate is the first gate in the LSTM cell that assists in forgetting or remembering information from the cell state. This is by applying a sigmoid function over the product of the forget gate weights W_f and the concatenated hidden state h_{t-1} and input x_t vectors. As the output of the sigmoid function σ gets closer to 1, keep the information. If the output is close to 0, then forget the information. For each value in the cell state vector, keep or discard its value according to the corresponding value returned from the forget gate.

$$f_t = \sigma\left(W_f \cdot [h_{t-1}, x_t] + b_f\right) \tag{8}$$

Figure 1-15 replaces the block representations of the three gates with mathematical symbols, omitting the weight symbols for simplicity.

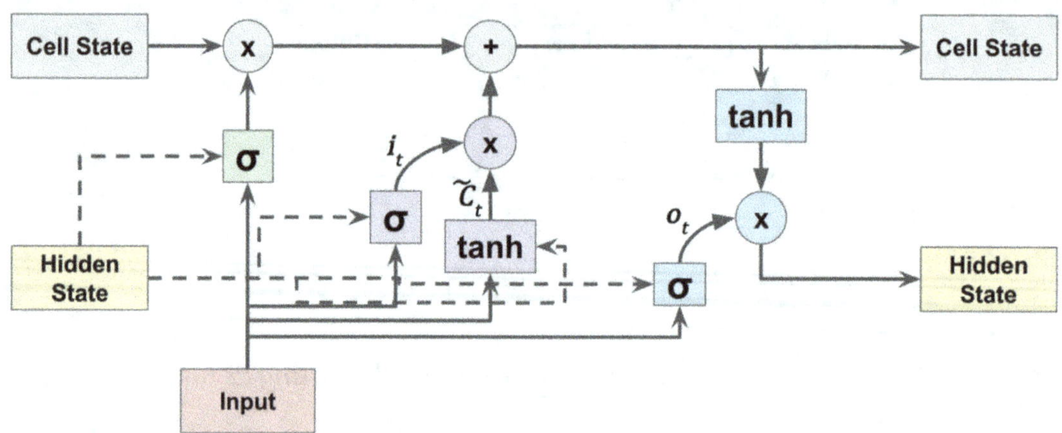

Figure 1-15. *Long short-term memory (LSTM) cell with mathematical symbols, showing the equations governing its input, forget, and output gates, as well as cell state updates*

1.7.2 Input Gate

The input gate decides what new information to store in the cell state. It creates two vectors:

1. A vector of a new potential cell state, denoted as $\tilde{C}_t$, computed based on the potential cell state weights W_c.

2. A vector of scaling values, denoted as i_t, computed based on the input gate weights W_i. It has values between 0 and 1 (using the sigmoid function), reflecting the amount of the new potential values to add to the original cell state.

The two vectors are multiplied to determine the amount of update to the cell state.

$$i_t = \sigma\left(W_i \cdot [h_{t-1}, x_t] + b_i\right)$$
$$\tilde{C}_t = \tanh\left(W_C \cdot [h_{t-1}, x_t] + b_C\right) \qquad (9)$$

By using the outputs from the forget and input gates, the new cell state is calculated.

$$C_t = f_t \cdot C_{t-1} + i_t \cdot \tilde{C}_t \qquad (10)$$

Remember that the issue with RNN is that the vanishing gradient problem appears due to the repetitive multiplications of small values returned by the sigmoid or tanh activation functions. If functions return values less than 1, then the gradients vanish. If they return values greater than 1, then the gradients explode.

1.7.3 Cell Gate

The cell state C_t does not depend on repetitive multiplications. It is just an almost linear equation without any activation functions. So, its gradient is almost constant and equal to 1. Even that C_t still depends on the forget gate output f_t, which in turn depends on the sigmoid activation function. The forget gate output is usually high, equal to 1, except for some situations where it produces a value close to 0 when the information is not needed. So, we can consider that the term $f_t. C_{t-1}$ is almost equal to C_{t-1}. Given that the other term $i_t.\tilde{C}_t$ is constant, then the equation can be considered as $C_t = C_{t-1}$, which does not depend on any repetitive multiplications or activation functions.

Note that the LSTM is not always resistant against vanishing gradients, as this problem could still happen if the forget gate output f_t is 0 for many steps (i.e., the input must be forgotten). But because this is not likely to happen all the time, LSTM is solving the vanishing gradient problem to some extent.

Finally, the output gate is responsible for measuring the amount of cell state to export as a hidden state at each time step. The cell state C_t is fed into the tanh activation function to produce a potential hidden state. This is then multiplied by the output of a sigmoid function o_t to determine the percentage of the potential hidden state to use as the new hidden state h_t.

$$o_t = \sigma\left(W_o \cdot \left[h_{t-1}, x_t\right] + b_o\right)$$
$$h_t = o_t \cdot \tanh\left(C_t\right) \tag{11}$$

The hidden state is dependent on the cell state. As long as the cell state remains resistant to the vanishing gradient problem, the hidden state will not be adversely affected.

Although the LSTM overcomes some of the limitations of the RNN, it still has some drawbacks:

1. LSTM mitigates the vanishing gradient problem, but it does not completely eliminate it.

2. LSTM cannot be efficiently parallelized.

3. While LSTM outperforms RNNs by capturing long-term dependencies through its memory mechanism, it struggles to handle extremely large sequences, such as those with thousands of tokens.

1.8 Gated Recurrent Unit (GRU)

The gated recurrent unit (GRU) is a more recent variation of the recurrent architecture, introduced as a streamlined alternative to the LSTM. While it maintains the ability to capture long-term dependencies and mitigate the vanishing gradient problem, it does so with a simpler internal structure.

The primary difference is that the GRU merges the LSTM's separate cell state and hidden state into a single hidden state, ht. Additionally, it reduces the number of gates from three to two:

1. The Reset Gate

2. The Update Gate

This reduction in complexity results in fewer parameters to train, often making GRUs faster to compute and more efficient on smaller datasets while maintaining performance comparable to LSTMs.

1.8.1 Reset Gate

The reset gate (rt) determines how much of the past information from the previous hidden state ht−1 should be forgotten. This is particularly useful for dropping information that is no longer relevant to the current part of the sequence. It calculates a value between 0 and 1 using a sigmoid activation function:

$$rt=\sigma(Wr \cdot [ht-1, xt]+br)$$

If rt is close to 0, the model effectively ignores the previous hidden state when calculating the new potential state, allowing it to start fresh if the sequence context shifts abruptly.

1.8.2 Update Gate

The update gate (zt) acts similarly to a combination of the LSTM's forget and input gates. It defines how much of the previous hidden state ht−1 should be kept and how much of the new candidate information should be added.

$$zt=\sigma(Wz\cdot[ht-1,xt]+bz)$$

A value of zt close to 1 means the model retains a large portion of the old memory, while a value closer to 0 allows the new candidate state to take over. This mechanism allows the GRU to remember long-term information for many time steps without the gradients vanishing.

1.8.3 Candidate Hidden State and Final Update

To calculate the final hidden state ht, the model first creates a candidate hidden state h̃t. This candidate uses the reset gate to scale the influence of the previous hidden state:

$$\tilde{h}t=\tanh(W\cdot[rt\cdot ht-1,xt]+b)$$

The final hidden state ht is then an interpolation between the previous state and the candidate state, controlled by the update gate:

$$ht=(1-zt)\cdot ht-1+zt\cdot\tilde{h}t$$

Because this final equation is a linear interpolation (similar to the LSTM's cell state update), it creates a shortcut for gradients to flow through time steps, significantly reducing the risk of vanishing gradients.

While the GRU offers a more computationally efficient alternative to the LSTM, it is not without its own set of limitations:

1. Because the GRU lacks a separate cell state and merges its memory into the hidden state, it has a slightly reduced capacity for storing complexity.

2. Like all recurrent architectures, the GRU must process tokens one by one. This prevents it from taking full advantage of modern parallel computing hardware, making training significantly slow.

3. Although the gating mechanism helps, the GRU still struggles to maintain coherence over very long sequences. As the sequence length grows into the thousands, the memory in the hidden state inevitably becomes diluted by more recent inputs.

4. Although the update gate significantly mitigates the vanishing gradient issue by allowing information to bypass multiple non-linear transformations, the GRU does not entirely eliminate vanishing gradients. In extremely deep sequences, the multiplicative nature of the backpropagation process can still cause gradients to decay, making it difficult for the model to learn relationships between very distant tokens.

Transformers address these issues by efficiently handling very long sequences while maintaining contextual awareness. Instead of relying on sequential connections, which can lead to vanishing or exploding gradients over time, transformers use the attention mechanism. This mechanism processes all tokens in parallel, eliminating the need for long chains of derivative multiplications.

Rather than treating all input tokens equally, attention mechanisms assign varying weights to emphasize the importance of each token in the context of the sequence. By prioritizing the most relevant tokens, attention mechanisms enhance the model's ability to establish meaningful mappings between input and output tokens.

CHAPTER 2

Transformers

Transformers extend traditional RNNs and LSTMs by introducing self-attention, the core mechanism behind their success. Unlike earlier models that process words sequentially, transformers assign varying attention weights to words based on their relevance to one another while encoding them into fixed-size vector representations.

The attention mechanism enables the model to focus on the most important words in a sentence rather than treating all words equally, mimicking human behavior in prioritizing key information. Moreover, self-attention can be computed in parallel, making transformers highly efficient and well-suited for GPU acceleration.

This chapter explains the encoder–decoder transformer architecture in detail, covering every step from encoding tokens into contextual embedding vectors to generating the predicted token using the decoder.

2.1 Encoder–Decoder Architecture

The encoder–decoder architecture is a fundamental framework in sequence-to-sequence (seq2seq) applications such as language translation, enabling the transformation of one sequence into another, potentially of different length. Both encoders and decoders are neural networks composed of dense, recurrent, convolutional, or other layers, depending on the problem's requirements. Figure 2-1 illustrates a generic architecture.

© Ahmed Fawzy Gad 2026
A. F. Gad, *Transformers and Large Language Models*, https://doi.org/10.1007/979-8-8688-2785-3_2

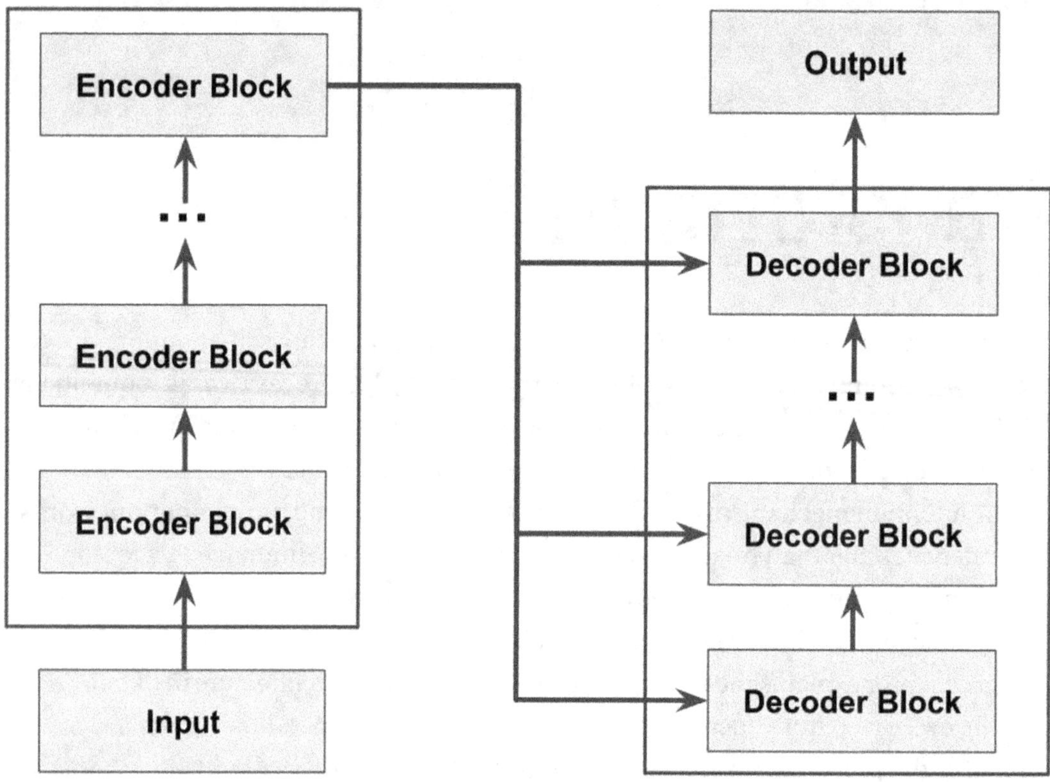

Figure 2-1. *Sequence-to-Sequence (Seq2Seq) architecture with encoding and decoding phases. The input sequence passes through multiple encoder blocks, producing an encoded representation that is fed into decoder blocks to generate the output sequence*

The encoding phase typically consists of multiple stacked encoder blocks, each following the same architectural design but with unique parameter sets. The blocks at the beginning focus on extracting low-level features, while deeper layers capture progressively higher-level abstractions. The greater the depth of the encoder, the more refined its contextual understanding of the input.

The output of the final encoder block, often referred to as the *context vector* or *encoded representation*, contains the distilled information from the entire input sequence. This representation is both compressed and context-rich, encoding the essential patterns, relationships, and structures present in the data. The encoded output can also be used as a feature vector for other downstream tasks, such as classification, clustering, or retrieval.

The decoding phase comprises multiple decoder blocks, each maintaining the same architecture but with distinct weights, ensuring an effective reconstruction or transformation of the encoded representation into the desired output sequence. The output of the last encoder block (feature vector) is fed as input to each block in the decoding phase. The encoder architecture can, for example, be an RNN that generates a hidden state, which is then passed to the decoder that could potentially be another RNN to accept such a hidden state and produce the output sequence.

Transformers are based on the encoder–decoder architecture, in which each encoder and decoder block incorporates an attention mechanism to dynamically refine embedding vectors according to their contextual relationships. This allows the model to capture dependencies between tokens regardless of their position in the sequence. Figure 2-2 illustrates the full architecture of the transformer model, showing the arrangement of both encoder and decoder components. The internal structure and function of each block will be examined in detail in subsequent sections.

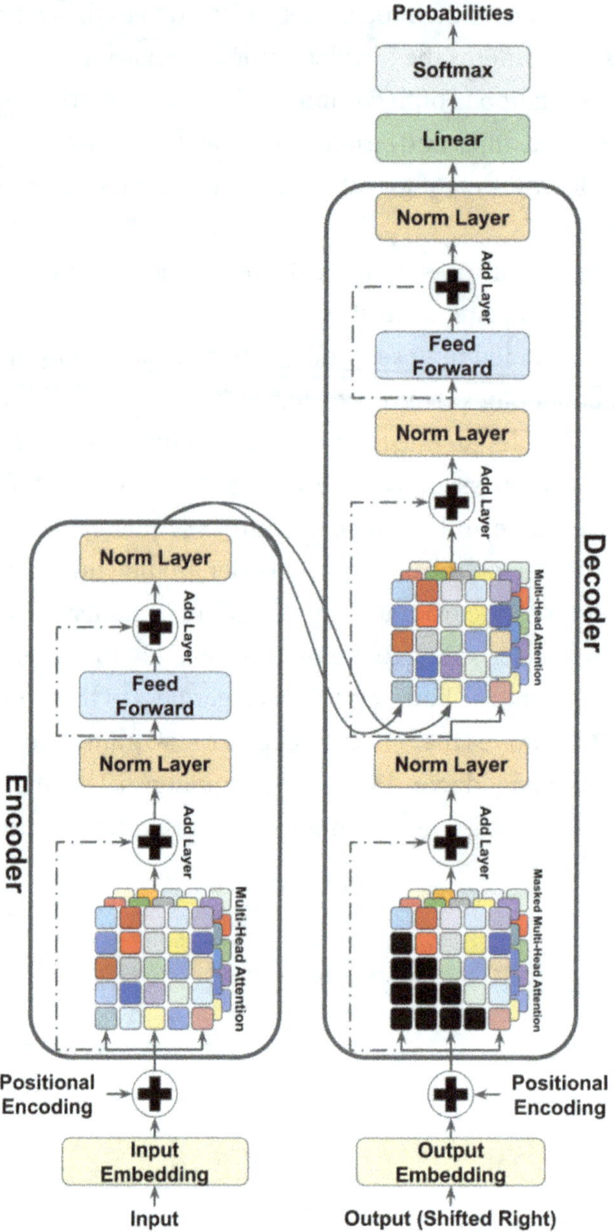

Figure 2-2. *Transformer model architecture, consisting of stacked encoder and decoder blocks. Each block integrates multi-head self-attention, feed-forward networks, residual connections, and layer normalization. The encoder processes the input sequence into contextual embeddings, while the decoder uses these embeddings, along with its own self-attention and cross-attention layers, to generate the output sequence*

One of the most popular transformer-based encoder models is BERT (Bidirectional Encoder Representations from Transformers), developed by Google in 2018. Other notable encoder models include RoBERTa (Robustly Optimized BERT Pretraining Approach) by Meta AI in 2019 and DeBERTa (Decoding-enhanced BERT with Disentangled Attention) by Microsoft in 2020.

For transformer-based decoder models, GPT-3 (Generative Pre-trained Transformer 3), developed by OpenAI in 2020, is a prominent example. Other decoder-style models include LLaMA (Large Language Model Meta AI) by Meta in 2023, Claude by Anthropic in 2023, and Gemini by Google in 2023 too.

There are also encoder–decoder (sequence-to-sequence) transformer models, such as T5 (Text-to-Text Transfer Transformer) by Google in 2020 and BART (Bidirectional and Auto-Regressive Transformers) by Meta AI in 2019, which combine both architectures to handle tasks like translation, summarization, and question answering.

2.2 Transformer Encoder

Generally, the transformer encoder consists of these steps as shown in Figure 2-2:

1. Input embedding

2. Positional encoding

3. Stack of N layers

    1. Multi-head self-attention layer

    2. Add & norm layer

    3. Feed-forward neural network (FFNN)

    4. Add & norm layer

The self-attention and dense (i.e., FFNN) layers are stacked N times. This is to capture more complex relationships and patterns in the input sequence.

Each of these two layers is followed by an element-wise addition layer and finally a normalization layer. Given the input embedding vector x, the outputs of such two layers can be represented as:

$$outputAttention = Norm\big(Add\big(x,Attention(x)\big)\big)$$
$$outputFFNN = Norm\big(Add\big(outputAttention,FFNN\big(outputAttention\big)\big)\big) \tag{1}$$

The next subsections discuss each of these steps.

2.2.1 Input Embedding

The first step in a transformer encoder is to generate embedding vectors. However, before this step, the text must first be tokenized using a technique such as WordPiece. Tokenizers rely on a predefined vocabulary that assigns a unique numeric ID to each token. The tokenized output consists of a sequence of these IDs, such as:

```
[12, 817, 174, ...]
```

Once tokens are converted into numeric IDs, they are mapped to embedding vectors using an embedding layer. The primary reason for embedding tokens as vectors is that machine learning models operate on numerical data rather than raw text.

Embedding vectors are dense, high-dimensional numerical representations of tokens. While these vectors are not interpretable, they encode meaningful relationships between words. Each dimension (or a combination of dimensions) captures specific linguistic properties.

Each element in an embedding vector can be thought of as capturing a specific property or feature of a word. While these features are not explicitly labeled, they emerge from the training process as the model learns meaningful representations. Below are some conceptual examples of how different dimensions in an embedding vector might encode various properties:

1. **Computing**: Phone, Laptop

2. **Fruit**: Apple, Orange

3. **Psychology**: Emotion, Mental

4. **Health**: Patient, Cancer

5. **Gender**: Male or Female

In reality, these properties are distributed across multiple dimensions rather than being isolated in a single element. The higher the quality of the embeddings, the better they capture complex relationships, allowing the model to understand contextual and semantic nuances more effectively.

Each word receives a score for each property, reflecting its degree of similarity to that property. Higher scores indicate a stronger association between the word and the given feature. Table 2-1 provides an illustrative example.

Table 2-1. *Scoring the words according to the degree of similarity*

	Computing	Fruit	Psychology	Health
Patient	0	0	0.5	1.0
Apple	0.7	1.0	0	0.3

In practice, an embedding vector can have thousands of dimensions, with each element representing a learned numerical feature. These features are complex and not directly interpretable, as there is no one-to-one mapping between a specific element and a specific property of the word. Instead, the meaning of a word emerges from the combined pattern across all elements.

Embedding vectors enable models to capture relationships between words by measuring distances or similarity scores in the embedding space. Words that frequently appear in similar contexts are positioned closer together in this space.

For example, embeddings may place the words *fruit* and *vegetable* near each other because they share strong semantic similarity. If two other words, say *orange* and *carrot*, are found at a similar relative distance, it may suggest that one is a type of fruit and the other a type of vegetable.

Tokens or words with similar meanings generally have similar embedding vectors. For example, if the embedding vector for the word *skip* is represented as:

```
[0.5, 7, 3, 12]
```

then the vector for the word *pass* is likely to be similar:

```
[0.48, 7, 3, 12.1]
```

Through training, the model learns associations between words based on their contextual usage, as in Figure 2-3. For instance, pairs such as *woman* and *man*, *king* and *queen*, or *apple* and *banana* tend to have similar vector representations. These relationships emerge because such words frequently appear in similar contexts within the training data.

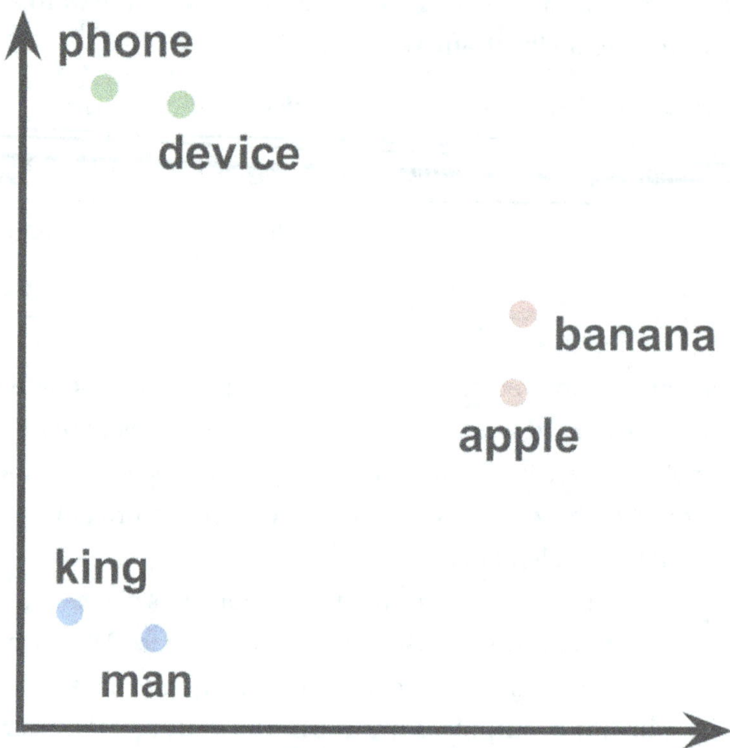

Figure 2-3. *Initial embedding space. Similar words such as king–man, banana–apple, and phone–device are located close to each other in the embedding space. Before applying contextual embeddings, words are positioned based on their general meaning. For example, an apple is placed near other fruits because it is understood as a type of fruit*

In the high-dimensional space of embedding vectors, semantically related words, such as *man* and *woman*, are positioned closer together compared to unrelated words like *device*. This structure also captures broader semantic relationships. For example, the distance between *sister* and *brother* may be comparable to the distance between *woman* and *man*, reflecting underlying gender-based associations.

Typically, each token is assigned a single embedding vector, capturing various aspects of its meaning. However, the interpretation of a token can be strongly influenced by its surrounding context. For example, the word *apple* may refer to either a fruit or a technology brand, depending on the sentence in which it appears.

This limitation is addressed by the attention mechanism, which allows models to refine an initial embedding vector by dynamically updating it based on the entire context of the input. This ensures that the representation of a word is adjusted to its meaning within a given sentence. Given the following input:

```
I have an apple iPhone
```

Given that the word *apple* is no longer referring to fruits within its context, it is moved closer to the word *phone* according to its context, as in Figure 2-4.

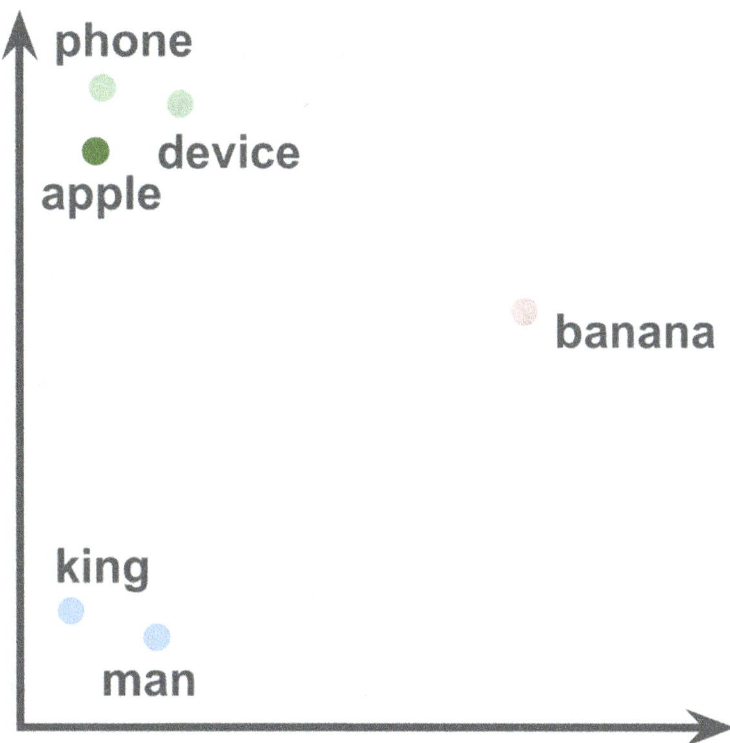

Figure 2-4. *Embedding space after updating the vector of the word apple according to its contextual meaning. In this context, Apple no longer refers to a fruit, so it moves away from banana and closer to device and phone in the embedding space*

Embedding vectors often have lengths in hundreds or thousands to capture a wide range of semantic properties. Table 2-2 summarizes several popular embedding models.

For example, the BERT-Base model produces embedding vectors of length 768 for each token. If the maximum tokenized sentence length is 512 tokens, the resulting embedding matrix for the sentence has a shape of *(512, 768)*.

In comparison, GPT-3 uses embedding vectors with 12,288 dimensions. Given that GPT-3 can process up to 2,048 tokens at a time, and each token is represented by a 12,288-dimensional vector, the resulting input matrix has a shape of *(2048, 12288)*. This matrix represents the total contextual information available to the model in a single forward pass.

Table 2-2. *Summary of popular transformer-based embedding models, showing their maximum token capacity, embedding size, number of hidden layers, number of attention heads, vocabulary size, tokenizer type, and release year*

Model	Max Tokens	Embedding Size	Hidden Layers	Attention Heads	Vocab Size	Tokenizer	Year
BERT-Base	512	768	12	12	30,522	WordPiece	2018
BERT-Large	512	1,024	24	16	30,522	WordPiece	2018
RoBERTa-Large	512	1,024	24	16	50,265	BPE	2019
DistilBERT	512	768	6	12	30,522	WordPiece	2019
ALBERT-Large	512	128	24	16	30,000	SentencePiece	2019
GPT-2	1,024	768	12	12	50,257	BPE	2019
GPT-3	2,048	12,288	96	96	50,257	BPE	2020
text-embedding-3-large	8,191	3,072	N/A	N/A	N/A	BPE	2024
XLNet-Large	512	1,024	24	16	32,000	SentencePiece	2019
Sentence-BERT	512	768	12	12	30,522	WordPiece	2019
MPNet-Base	512	768	12	12	30,522	WordPiece	2020
slate-125m-english-rtrvr	512	768	12	12	30,000	SentencePiece	2024

(continued)

Table 2-2. (*continued*)

Model	Max Tokens	Embedding Size	Hidden Layers	Attention Heads	Vocab Size	Tokenizer	Year
text-embedding-005	8,191	1,536	N/A	N/A	50,257	BPE	2024
Linq-Embed-Mistral	32,768	4,096	32	32	32,000	BPE	2024
Cohere-embed-english-v3.0	512	1,024	12	16	256,000	Custom	2023
bilingual-embedding-large	512	1,024	12	16	50,000	BPE	2024
jina-embeddings-v3	8,191	1,024	12	16	32,000	BPE	2024
Snowflake-Arctic-embed-l	512	1,024	12	16	50,000	BPE	2024
voyage-3-large	32,000	1,024 (default), 256, 512, 2,048	N/A	N/A	N/A	N/A	2025

Such embeddings are stored in an embedding matrix that is initially assigned random values and then learned during the training process. In the BERT-Base model, there are 30,522 tokens. Given that its embedding size is 768, then the embedding matrix size is *30,522x768*.

2.2.2 Positional Encoding

The positions of tokens are crucial for accurately understanding text. NLP models must recognize the position of each token within a sentence to preserve meaning. Without positional awareness, a model would treat sentences containing the same words as identical, regardless of their order.

For example, consider the following sentences:

1.   The player kicks the football.

2.   The football kicks the player.

Although both sentences contain the same words, their meanings are entirely different due to the word order.

In RNNs and LSTM networks, positional information is inherently preserved due to their sequential nature. The model processes tokens in order, with the first input always corresponding to the first token, followed by the second token, and so on, until the final token is processed.

Similarly, in CNNs, the convolution operation maintains positional awareness by processing blocks of pixels. The operation begins at the top-left corner and progresses sequentially to the bottom-right corner, ensuring spatial relationships are preserved.

Transformers rely on the attention mechanism to capture the relationships between words within a given context. Unlike RNNs and CNNs, the attention mechanism does not inherently encode positional information, as it lacks recurrence and convolution.

Using only attention, the model can identify a relationship between words such as *player* and *football*. However, without positional encoding, it would not distinguish between different word orders, which are crucial for understanding the sentence's meaning.

To address this challenge, the model must be aware of the position of each token. In transformers, this is addressed using positional encoding, allowing the model to recognize the position of each token as in Figure 2-5. The positions are encoded into the embedding vectors outside the attention block before the encoder or decoder blocks.

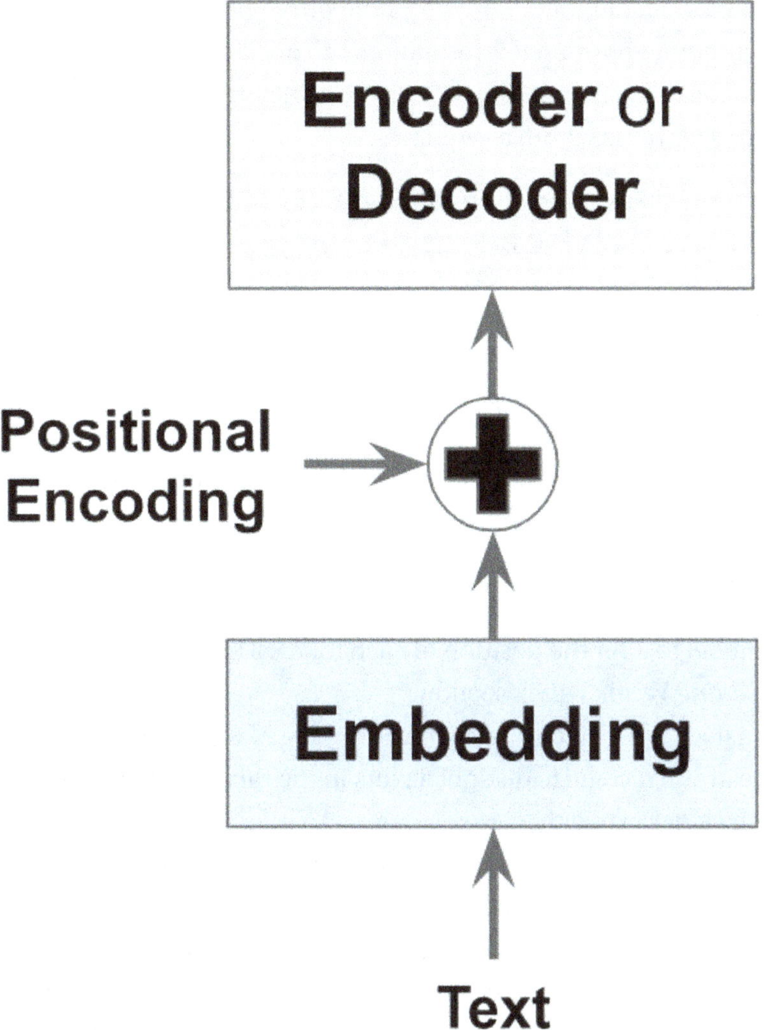

Figure 2-5. *Positional encoding adds position information to the embedding vector, enabling the model to recognize the position of each token in the sequence*

In transformer models, positional encoding is represented as a vector of the same length as the embedding vector. For example, if the embedding vector has a length of 2048, the corresponding positional encoding vector also has a length of 2048.

The positional encoding vector is computed once and applied to all embedding vectors. To incorporate positional information, the positional encoding vector is added elementwise to the token embedding vector, ensuring that the model can distinguish the relative positions of words within a sequence.

```
EmbeddingVectorwithPosition = EmbeddingVector + PositonalEncodingVector
[ep0, ep1, ep2, ..., epN] = [e0, e1, e2, ..., eN] + [p0, p1, p2, ..., pN]
```

This raises an important question: **How does the model differentiate between the embedding vector value and the position value?** If an embedding vector value is 1.4 and the encoded position value is 0.5, their sum is 1.9. How does the model know that the position component was originally 0.5?

The key is that the model does not need to separate the embedding and position values after summation. Instead, it learns to use the combined values during training. The structured pattern of positional encodings persists even after being added to the embeddings. Neural networks are remarkably effective at recognizing such patterns, allowing the model to infer the position of each token. This will become clearer in the *Position Embedding Vector Values* section.

A useful analogy is a fruit cocktail: when drinking it, we don't separate each fruit, but our brain can still recognize distinct flavors in the mix. Similarly, the model doesn't extract position values explicitly.

This ability to uncover hidden patterns is a fundamental strength of deep learning. A well-known example comes from computer vision: early models relied on hand-crafted features to identify objects like cars in images, which was challenging to find a set of features able to identify such objects in different conditions. Deep learning models learn such distinguishing features directly from data by exploring the identifying patterns in the objects. Even if these learned features aren't always interpretable, they work remarkably well in practice.

An important consideration in positional encoding is that its values must not be too large to disrupt the similarity between word embedding vectors. At the same time, they must not be too small, or their effect will become negligible.

For example, suppose the similarity score between the words *king* and *queen* is 0.9 before adding positional encoding. After adding positional encodings, this similarity should remain close to 0.9, ensuring that positional information does not overpower semantic relationships.

Consider the following simplified word embedding vectors:

1. king: $[1, 2]$

2. queen: $[1, 2.2]$

If the similarity is measured using the dot product, then:

```
king.queen=[1, 2].[1, 2.2] = 1 + 4.4 = 5.4
```

Now, assume the positional encoding vector is $[0.05, 0.02]$. After adding positional encoding, the updated embeddings become:

1. king: $[1.05, 2.02]$

2. queen: $[1.05, 2.22]$

The new similarity score is:

```
1.1025 + 4.4844 = 5.5869
```

This new similarity score of 5.5869 is close to the original 5.4. Figure 2-6 illustrates that the distance between the two vectors remains nearly the same before and after adding positional embeddings, ensuring that word similarities are preserved.

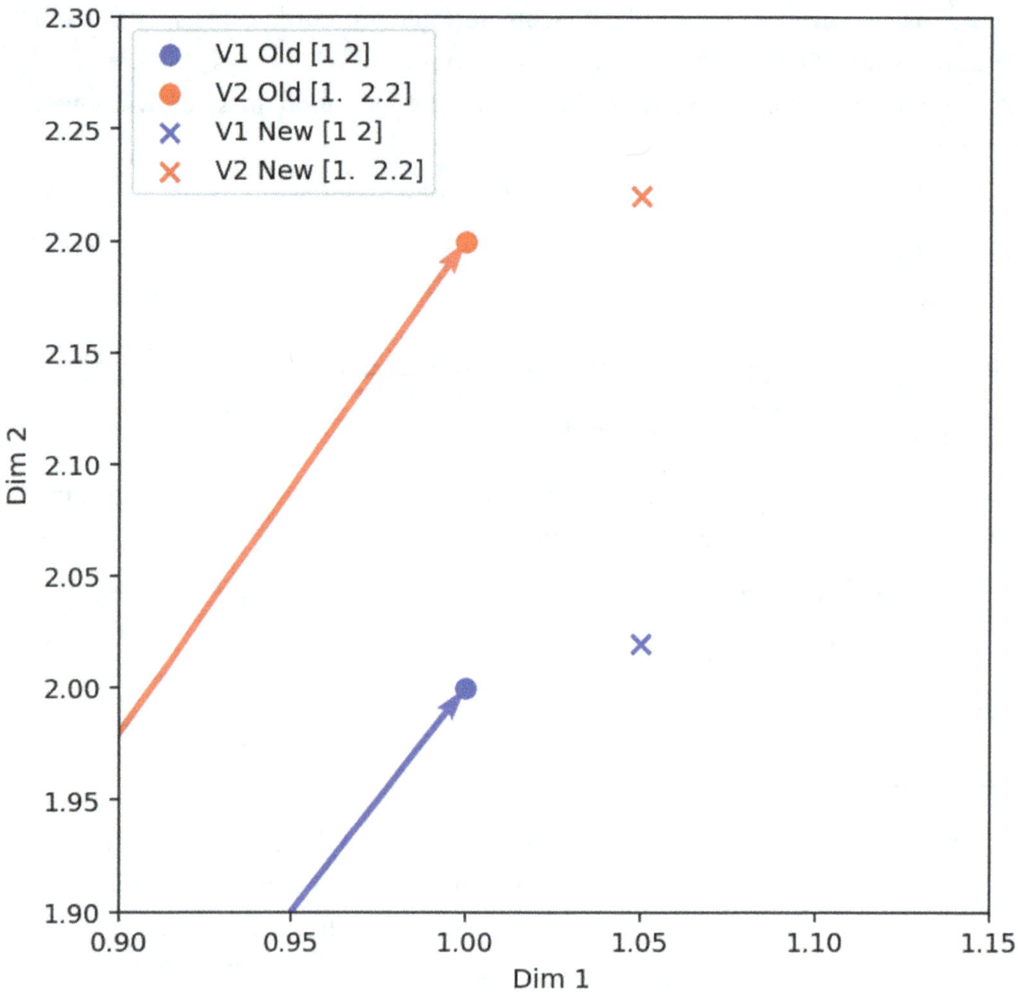

Figure 2-6. *The embedding vector before and after adding the positional encoding vector. The distance between the updated vectors changes only slightly, ensuring that the similarity between tokens remains nearly the same after positional embedding*

Positional embeddings should only modify word embeddings without distorting their core meaning or altering the similarity between two tokens. The similarity or the distance should remain nearly the same as it is not the responsibility of positional embeddings to alter the token similarity or move the vectors into the embedding space. This is the responsibility of the attention mechanism. Instead, positional embeddings adjust the positions of word embeddings within their high-dimensional space to encode word order while preserving semantic relationships.

Sinusoidal Positional Embedding

Now that the concept of positional embeddings is clear, the next question is, how do we generate the encoded position values?

An ideal function for generating positional encodings must satisfy the following criteria:

1. The generated values should not be too large, ensuring that positional encodings do not overshadow or distort the actual word embedding vectors.

2. The function must remain effective even as the sequence length increases, making it adaptable for models with varying token limits (i.e., different context sizes).

3. Each position must have a distinct positional encoding vector to differentiate tokens at different locations in a sentence. If two tokens had identical positional encodings, the model would lose positional information.

Common activation functions like sigmoid or ReLU (rectified linear unit) are not suitable for this purpose. Their output ranges are limited, and they tend to saturate by producing the same values beyond a certain input range. For example, a ReLU function might output a constant value (e.g., 1.0) for all large inputs, making it ineffective for encoding unique positions.

Sine and cosine functions are well-suited for generating positional encodings because they satisfy the key requirements:

1. Their outputs range from -1 to +1, ensuring they do not overpower or distort the embedding vector.

2. As periodic functions, sine and cosine waves can generate values for infinitely long sequences without saturating, making them effective regardless of sequence length.

3. A challenge with sine and cosine waves is that they naturally repeat over time. Depending on the wave frequency, different input positions could map to the same output values, losing uniqueness.

To prevent repetition, the frequency of the sine and cosine waves must be carefully adjusted. By selecting appropriate frequency scaling, the functions can generate unique positional encodings even for very long sequences. This is a critical consideration when designing transformer architectures.

The values within the positional encoding vector are generated by sampling sine and cosine waves, as defined by the following equations. Elements at even indices are derived from a sine wave, while those at odd indices are sampled from a cosine wave. This ensures a structured pattern and a unique positional representation for each token in the sequence. Creating such a pattern is important so that the neural network can capture it and deduce the positions.

$$PE_{(pos,2i)} = \sin\left(\frac{pos}{10000^{\frac{2i}{d}}}\right)$$

$$PE_{(pos,2i+1)} = \cos\left(\frac{pos}{10000^{\frac{2i}{d}}}\right)$$

$$(2)$$

The variable i represents the index of each element in the embedding vector, pos variable is the position index of each token in the input sequence, and d is the embedding dimension. If the embedding vector has a length of L, then i ranges from 0 to $\frac{L}{2}$.

For instance, if the embedding vector length is $L = 512$, a pair of sine and cosine values is generated for each index from 0 to 255, where sine is applied to even indices (0, 2, 4, ..., 254) and cosine to odd indices (1, 3, 5, ..., 255). This results in a total of 512 values, ensuring that each element in the embedding vector receives a unique positional encoding.

The frequency of each sine and cosine wave is determined by:

1.   The token's position in the sequence.

2.   The element index within the embedding vector.

This ensures that each token and each element in its embedding vector receives a unique wave with a distinct frequency, preventing repetition and preserving positional information effectively.

At each position i within the embedding vector, the sine and cosine functions are applied. The $2i$ and $2i + 1$ terms are used to produce two different indices out of each i value, one for sine and another for cosine, along the range from 0 to $\frac{L}{2}$. Specifically:

- For index $i = 0$, the pair of values is $PE(pos, 0)$ from the sine wave and $PE(pos, 1)$ from the cosine wave.

- For index $i = 1$, the pair is $PE(pos, 2)$ and $PE(pos, 3)$.

This alternating assignment continues throughout the embedding vector, ensuring each position receives a unique combination of sine and cosine values.

Figure 2-7 illustrates sine and cosine waves used for positional encoding when the input sequence consists of 100 positions (tokens). Each subplot represents the waveforms that generate positional values, shown here for only two dimensions of an embedding vector with a total length of 16.

As the position index i increases, the denominator in the sine and cosine functions grows, leading to a decrease in frequency. This explains why the waves have higher frequencies near index 0 and gradually smooth out as they approach the upper limit of the embedding dimensions.

The advantages of the frequency variation include:

1. Each position has a unique but smoothly varying encoding.

2. The model can differentiate positions and capture their patterns while maintaining relational structure.

3. Reducing the chance of collision where different positions might have the same positional encoding value.

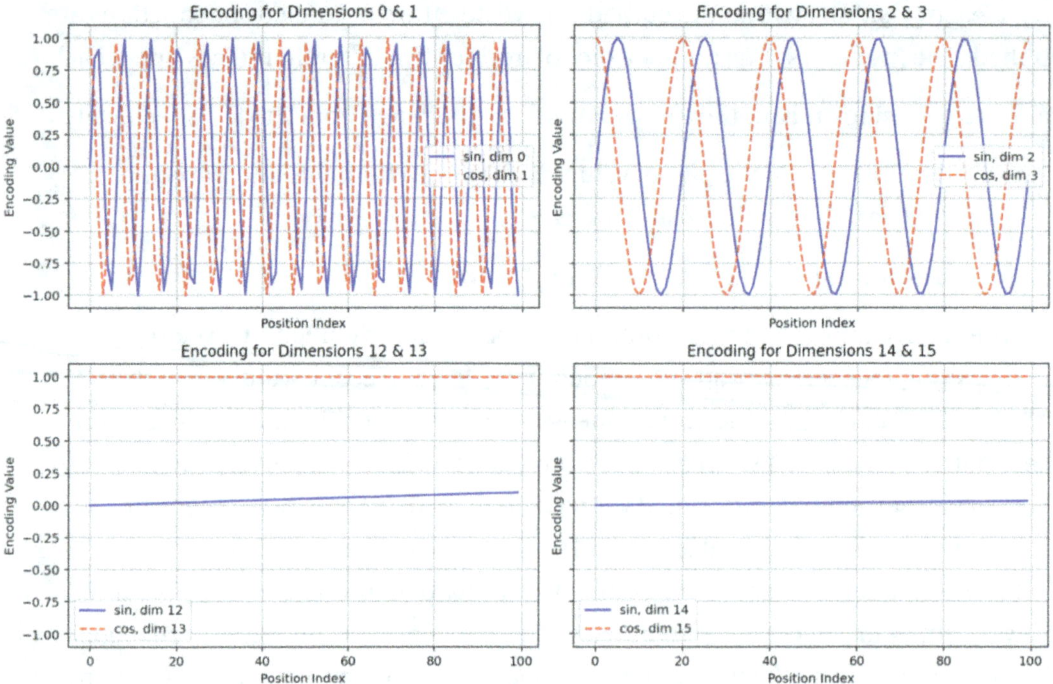

Figure 2-7. *Sine and cosine waves are used for positional encoding for an input sequence of 100 tokens. The example shows two dimensions of a 16-dimensional embedding vector. As the position index increases, the wave frequency decreases, producing unique yet smoothly varying encodings that help the model distinguish positions, preserve relational structure, and reduce the chance of collisions*

Figure 2-8 illustrates the values in the positional encoding vectors of length 16 for multiple positions when the input sequence consists of 100 tokens. Each position has a unique pattern of positional encoding values. Neural networks are highly effective at recognizing such patterns. Even after summing up the positional encoding vector with the embedding vector to form a position-aware embedding, the network can still distinguish positional information and infer the token position from the modified embedding vector.

To better illustrate the variations, the constant 10,000 is replaced with 100 when generating the values in Figure 2-8. This increases the frequency of the waves, making the patterns more noticeable, especially since this example uses a small embedding vector of only 16 dimensions.

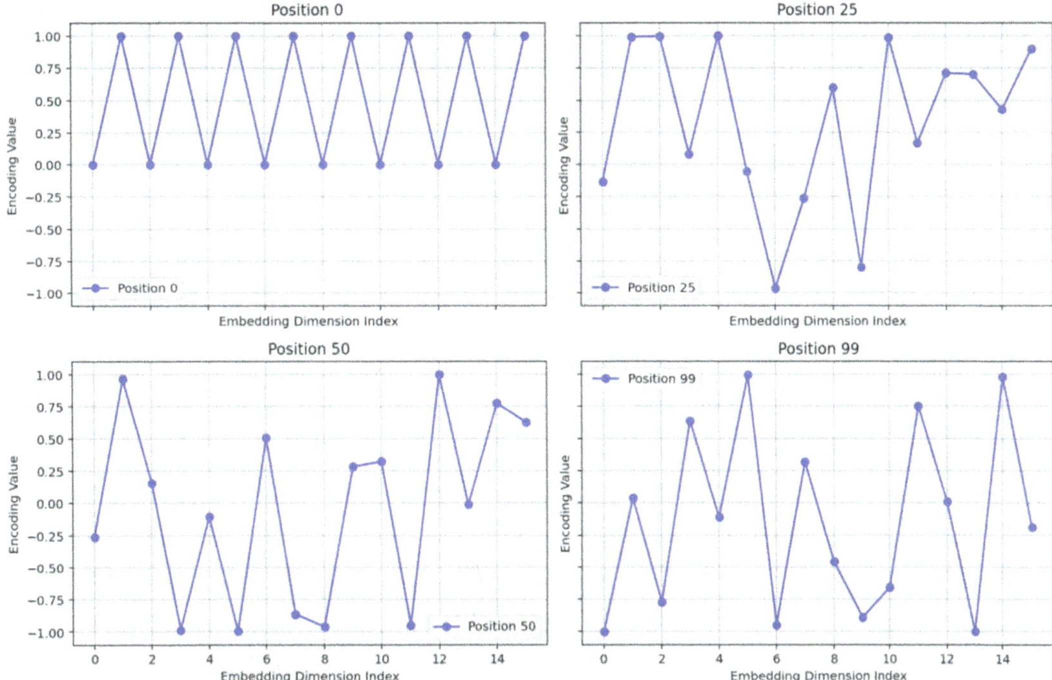

Figure 2-8. *Positional encoding vectors of length 16 for multiple positions in a 100-token sequence. Each position exhibits a unique pattern that neural networks can readily recognize. Even after being added to the embedding vector, these patterns preserve positional information, allowing the network to infer token positions. For clarity, the constant 10,000 is replaced with 100 to increase wave frequency and highlight variations in this small 16-dimensional example*

During training, the encoder focuses on learning improved embeddings for each token using the attention mechanism. This is by moving the vector in the embedding space to be closer to other tokens based on the context. These embeddings (i.e., learned features) are then passed to the decoder for further processing.

Why Not Concatenation?

An alternative to summing up the position and embedding vectors is to concatenate them, preserving both the embedding values and the encoded position. However, this approach presents several challenges.

```
[e0, e1, e2, ..., eN, p0, p1, p2, ..., pN]
```

The primary issue is the significant increase in the number of trainable parameters. If only the embedding vector is used without concatenation, the number of parameters required for the Query Q, Key K, and Value V weight matrices is $128 \times 12{,}288 = 1{,}572{,}864$ (for GPT-3). This is because GPT-3 has an embedding size of 12, 288 distributed across the 96 heads, resulting in $d_k = d_v = \dfrac{12{,}288}{96} = 128$.

If a positional encoding vector is introduced to capture positional information for each token, and its length matches that of the embedding vector (12,288), concatenating the two results in a new vector of length $12{,}288 \times 2 = 24{,}576$. Consequently, the Key/Query/Value weight matrix size doubles to $128 \times 24{,}576 = 3{,}145{,}728$, leading to increased memory requirements and longer training times.

One possible solution is to use a smaller positional encoding vector rather than matching the embedding vector's length. For instance, if the embedding vector length is 12,288, the positional encoding vector could be reduced to 512. However, this introduces an additional hyperparameter that must be fine-tuned.

Another important question is: *Why not simply learn the positional vectors instead of using predefined ones?* While this is possible, it introduces a significant drawback. Transformer architectures already contain a large number of parameters, and a key design objective is to minimize the model parameter count to improve efficiency. Learning positional vectors would effectively double the number of trainable parameters related to input representations, leading to increased memory usage and training costs. As a result, using sinusoidal functions (sine and cosine) to generate positional encodings remains a more efficient and practical alternative.

Although having a single vector represents both the token embedding and its position may seem like a black box, it is an effective approach. Results demonstrate that the model successfully captures positional information, proving its awareness of token positions.

2.2.3 Encoder Block Internals

So far, the input text has been tokenized, each token has been represented as an embedding vector, and positional encoding has been applied to embed positional information into the vector. However, each token is still represented by the same embedding vector regardless of external factors such as position or context.

A single word can have different meanings depending on the surrounding tokens. For example, the words *light* and *model* in the examples below take on different meanings depending on their context. Despite this, their embedding vectors remain the same, irrespective of the words around them.

- *Turn on the light.*

- *This bag is light.*

- *A machine learning model ...*

- *A fashion model ...*

This limitation means that static word embeddings fail to capture contextual meaning, which is crucial for accurate language understanding. This indicates how important it is for the model to understand the relationship between each token and its context.

Remember that one of the key challenges in RNN and LSTM models is the vanishing gradient problem. As the sequence length increases, the gradients become progressively smaller, preventing distant cells from effectively learning. Consequently, RNNs and LSTMs struggle to capture long-range dependencies between words in a sentence.

In contrast, CNNs are limited by their receptive field. They can only establish relationships within the area covered by their convolutional kernels. For example, if the kernel size is 3×3, the model can only capture patterns within that block and cannot directly relate to elements outside of it.

The encoder block in the Transformer model architecture addresses this issue by utilizing the self-attention mechanism. This mechanism determines how much each token in the input sequence attends to other tokens, as in Figure 2-9.

Based on these attention scores, each token embedding vector is updated by incorporating contextual information from the entire text sequence. For example, if the scores indicate a strong relationship between the tokens "began" and "ended," their embedding vectors will be updated accordingly. Suppose we are updating the embedding vector of the word "began" based on its attention score with "ended," which is 0.8. Given the embedding vectors of these two tokens, the update will reflect their contextual relationship.

$$E_{began} = [0.1, 0.3, -0.5]$$
$$E_{ended} = [0.81, -0.04, -0.7]$$
$$Enew_{began} = E_{began} + 0.8 * E_{ended} \qquad (3)$$
$$= [0.1, 0.3, -0.5] + [0.648, -0.032, -0.56]$$
$$= [0.748, 0.268, -1.06]$$

As a result, the same word can have different representations depending on its surrounding context, enabling the model to dynamically capture word meanings more effectively.

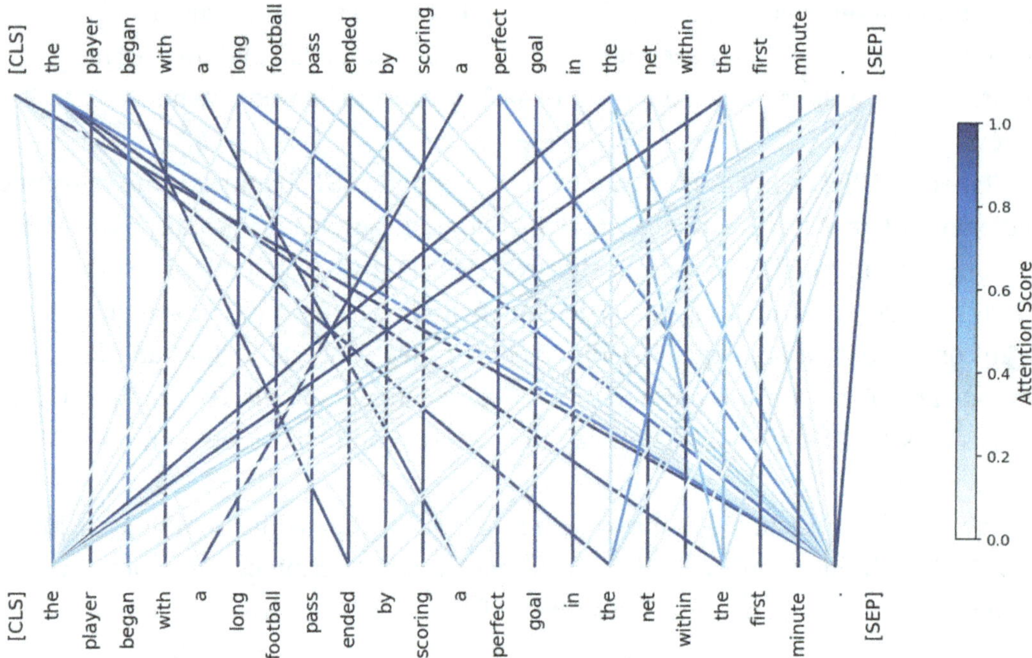

Figure 2-9. *Scores from the self-attention mechanism. Attention scores determine how much each token influences others in the sequence*

It is common for models to apply multiple attention functions, known as heads, allowing the model to capture different aspects of word meaning. For example, using eight parallel attention heads, the model can extract eight different contextual meanings for the same token by applying self-attention according to the number of heads.

Self-attention uses a similarity metric to measure how similar two embedding vectors are. Some of the commonly used metrics are:

1. Scaled-dot product

2. Cosine

3. Additive

2.2.4 Scaled Dot-Product Attention

Transformer-based embedding models begin with generic embedding vectors assigned to tokens based on a predefined dictionary. Initially, a token always receives the same embedding vector regardless of its context. For example, in the following three sentences, the word *model* has different meanings, yet its initial embedding remains identical.

```
A machine learning model is being trained.
A fashion model is on the stage.
Please model your plan.
```

The generic embedding vector for *model* is the same across all three sentences.

```
[0.7949, 0.5223, 0.0392, ..., 0.0762]
```

Next, positional encoding is applied, modifying the embedding vectors based on the token position in the sentence. While these new vectors remain almost close to the original, they now encode positional information.

```
[0.7949, 0.5023, 0.1392, ..., 0.0712]
[0.7939, 0.5253, 0.0592, ..., 0.0862]
[0.7749, 0.5123, 0.0982, ..., 0.0681]
```

After positional encoding, next is the scaled dot-product attention as in Figure 2-10. It is applied to refine the embedding vectors based on the surrounding context. As a result, the embeddings of *model* become more distinct in each sentence. In the high-dimensional space (e.g., 12,288 dimensions in GPT-3), this refinement is like moving the embedding vector so that it aligns more closely with relevant words in the context.

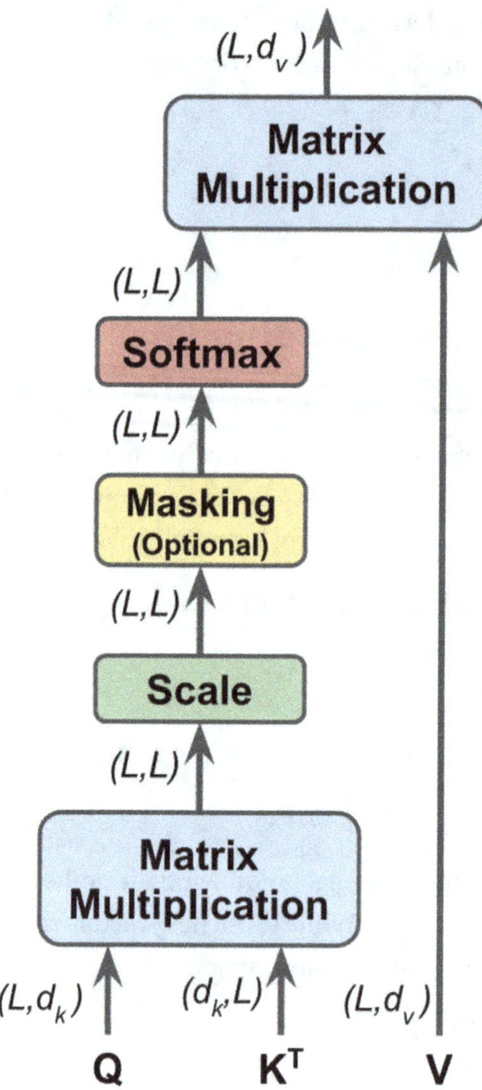

Figure 2-10. *Scaled dot-product attention. This mechanism computes attention scores by taking the dot product of query and key vectors, scales the result to prevent extreme values, and applies a softmax function whose output is multiplied by the value vectors*

For instance, the embedding of *apple* might shift toward *technology* in one case and toward *fruit* in another.

```
[1.2134, 0.4013, 0.1392, ..., 0.0361]
[0.2939, 0.5051, 0.4592, ..., 0.0742]
[0.8749, 0.5453, 1.2982, ..., 0.8183]
```

Before explaining how scaled dot-product attention works, let's quickly refer to information retrieval systems. Suppose we are searching for a video using the following query:

```
How to use the notes app on PC?
```

Each video in the database has a title or description, which we can refer to as a key. The search engine compares the user's query against these keys, measures similarity, and retrieves the most relevant videos (values) based on a similarity metric. A similar concept is applied for transformers.

Query, Key, and Value Matrices

For each embedding vector in the context, query, key, and value vectors are generated. All query vectors are grouped into the query matrix Q, while the key and value vectors are assembled into the key matrix K and value matrix V, respectively. The length of the query and key vectors is denoted as d_k, while the length of the value vectors is d_v. Although d_v is sometimes equal to d_k, this is not always the case.

The query, key, and value matrices are obtained by performing matrix multiplication between the embedding matrix E and the following weight matrices where their parameters are trainable:

1. W_q: Query weight matrix.

2. W_k: Key weight matrix.

3. W_v: Value weight matrix.

Suppose the embedding size is $d_{model} = 768$. Typically, the dimensions of the query, key, and value vectors, denoted as d_k and d_v, are set equal to the embedding size when using a single head of self-attention. Consequently, each of the weight matrices query W_q, key W_k, and value W_v has a size of $d_{model} \times d_k$. Since $d_k = d_{model}$ in single-head attention, the weight matrices are commonly represented as $d_{model} \times d_{model}$. However, expressing the size as $d_{model} \times d_k$ is preferable because it accommodates scenarios where d_k differs from d_{model}, such as in multi-head attention.

By multiplying the embedding matrix E of size $L \times d_{model}$ (where L is the maximum number of tokens in the input sequence) with these weight matrices, we obtain the query matrix Q, key matrix K, and value matrix V. If the context has $L = 2,048$ tokens, then the size of each of the 3 matrices is $L \times d_k = 2,048 \times 768$.

$$
\begin{aligned}
Q &= E \times W_q \\
K &= E \times W_k \\
V &= E \times W_v
\end{aligned}
\tag{4}
$$

The query W_q and key W_k weight matrices transform the embedding vectors into new spaces, making similarity measurements more effective.

A record in the query matrix Q can be thought of as a question about certain properties or features of the token. The purpose of these questions is to gain a clearer understanding of the token's surrounding context. While the exact questions represented by each query are not explicitly known, they can be imagined as inquiries such as:

1. Is there an adjective preceding the token?

2. Is there a verb preceding the token?

For example, the presence of an adjective before a word can alter its meaning. Consider the word *tree*, whose meaning is influenced by the preceding word:

```
a long tree ...
a thin tree ...
```

In both cases, the adjective modifies the perception of the tree, demonstrating how context impacts meaning.

While the records in Q represent questions, the records in the key matrix K can be thought of as answering such questions. When the query matrix is multiplied by the key matrix, the resulting scores indicate how strongly the token associated with the query relates to the query (i.e., question).

Imagine a record in the key matrix that measures the *roundness* of a token, encoded as a value ranging from 0 to 1, as in Figure 2-11. Tokens representing rounded objects, such as *apple, circle,* and *egg,* would have roundness scores close to 1 because they have this feature, while less round tokens like *box* and *American ball* would have scores close to 0. Similarly, other records in the key matrix measure dozens of additional features in the tokens.

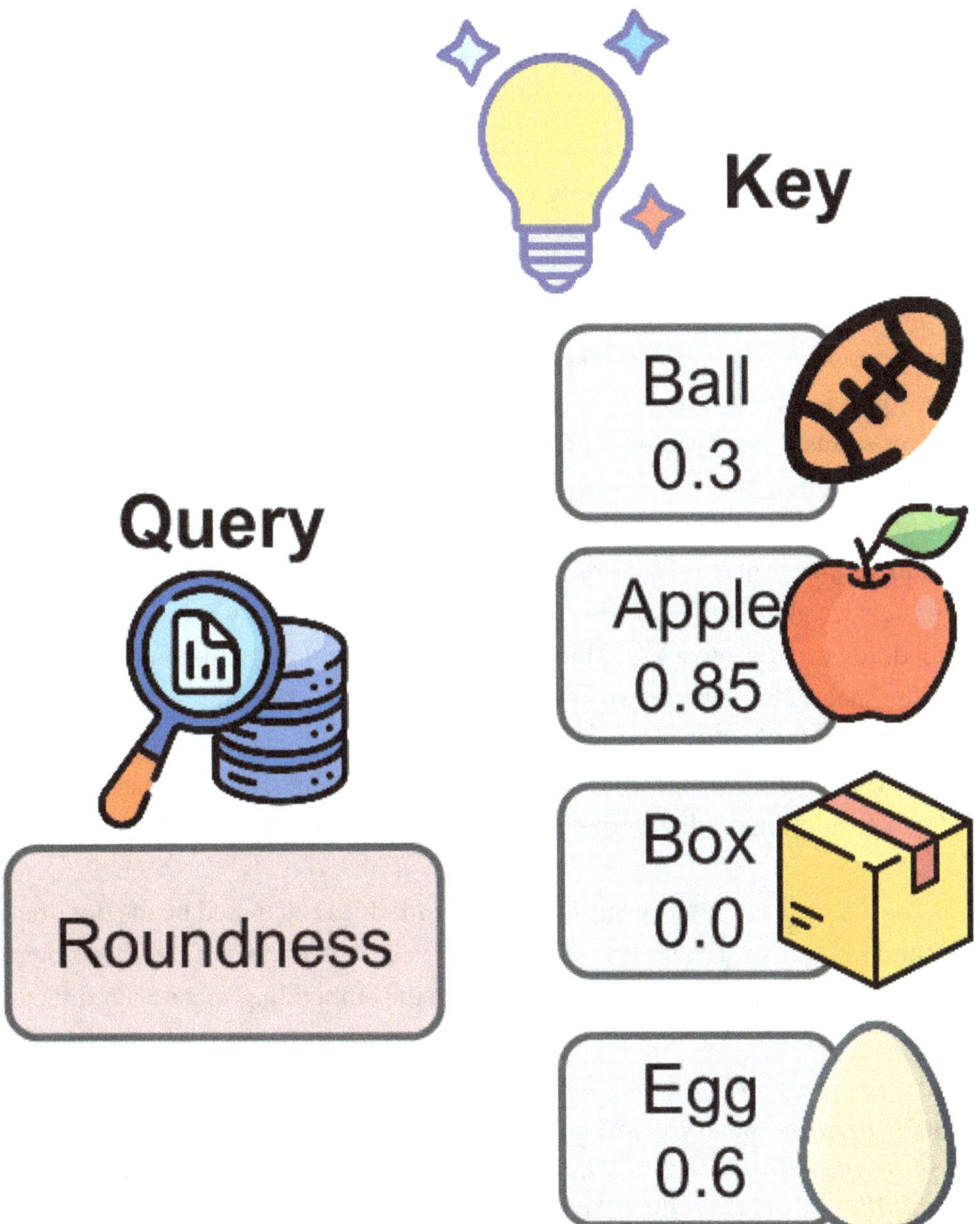

Figure 2-11. *The query represents a question, such as "What is the roundness?" The key stores features' measurements like roundness, with values ranging from 0 to 1. The apple and egg tokens have high scores, while less round tokens like the box and American ball have scores near 0*

During training, the model learns the optimal parameters in W_q and W_k to transform the embedding vectors into a new space optimized for finding similarities between tokens.

To understand this, consider an example of object matching in images. Suppose we need to measure the similarity between objects in two images. Using the raw pixel values directly would yield low accuracy because pixels are highly sensitive to variations in illumination, rotation, scale, and other transformations. Even if both images contain the same object at different scales, the direct pixel comparison would likely fail.

A better approach is to extract features such as SIFT (Scale-Invariant Feature Transform), which are invariant to such changes, and use them to match the images. This process transforms the images from one space (raw pixels) into another space (feature descriptors), making it easier to compare objects and measure their similarity. Similarly, the query and key matrices in attention mechanisms transform the embedding vectors into a space where measuring similarity between tokens becomes more effective.

The value weight matrix W_v is used to create the value matrix V, which transforms the embedding vectors into a space better suited for repositioning them within the embedding space. The value matrix reflects the amount of information that should update each token to be better aligned with the other attending tokens.

This transformation brings similar tokens, based on their context, closer to each other. Such positioning facilitates the prediction of the next token, as in Figures 2-3 and 2-4.

The value matrix is scaled according to the similarities calculated by the dot product of the Q and K matrices. The scaled V matrix is then added to the input embeddings for the purpose of moving the embedding vector into the embedding space according to its context.

The space of the Q and K matrices cannot assist in predicting the next token. This is because the embedding created by these 2 matrices is optimized to capture the features (e.g. color, technology, fruit, food, etc) and calculating the similarity between the tokens.

The space of the V matrix is optimized to create embeddings that measure if 2 words are in the same context. For example, it knows that the words car and vehicle are similar.

Dot Product Similarity

The three matrices Q, K, and V are used to compute scaled dot-product attention between pairs of tokens in the input sequence using the following equation.

$$\text{Attention}(Q,K,V) = \text{softmax}\left(\frac{QK^{\top}}{\sqrt{d_k}}\right)V \tag{5}$$

For each embedding vector (i.e., token), three new vectors are calculated:

1. **Query Q_i:** It uses the query weight matrix W_q.

2. **Key K_i:** It uses the key weight matrix W_k.

3. **Value V_i:** It uses the value weight matrix W_v.

The multiplication of the query vector Q_i and the key vector K_i represent the similarity between tokens. Note that the keys matrix must be transposed before applying the matrix multiplication. Based on this similarity, the value vector V_i is updated to reposition the token embedding vector according to its context.

A matrix is created out of the dot product between each key-query pair representing the similarities. The product shows how strong the relationship between the tokens is. The matrix shape is $L \times L$.

If the model can handle, as in GPT-3, 2, 048 tokens at a time, then there are 2, 048 embedding vectors. As a result, the size of the matrix is 2, 048 × 2, 048.

$$S = QK^{\top} = \begin{bmatrix} q_1 \cdot k_1 & q_1 \cdot k_2 & \cdots & q_1 \cdot k_n \\ q_2 \cdot k_1 & q_2 \cdot k_2 & \cdots & q_2 \cdot k_n \\ \vdots & \vdots & \ddots & \vdots \\ q_m \cdot k_1 & q_m \cdot k_2 & \cdots & q_m \cdot k_n \end{bmatrix} \tag{6}$$

When the query q_i and the transposed key k_i^T vectors match, then their dot product becomes large positive numbers compared to those that do not match. The numbers range from $-inf$ to $+inf$ depending on how similar the vectors are.

The dot product produces a similarity score, as shown in Table 2-3, between the key and query vectors, indicating how relevant each word is in updating the meaning of the other word. The higher the similarity between the two tokens, the greater their dot product value.

Table 2-3. *Dummy dot-product matrix*

	T1	T2	T3	T4	T5	T6
T1	12	8	7	10	17	15
T2	9	9	2	9	16	9
T3	11	7	1	14	9	12
T4	10	6	5	1	12	3
T5	11	0	3	6	18	8
T6	1	3	5	2	6	5

Note that the dot product is only a way of measuring similarity. Other similarity metrics can also be used, like the cosine similarity. This way we are measuring an angle from the origin of the embedding vector to reflect the similarity. As the cosine ranges from -1 to +1, then the tokens are more similar when they are close to +1. The dot product and cosine similarity are closely related when the embedding vectors are normalized to have a unit norm $\|x\| = 1$.

Scaling Factor

In the attention equation, each dot product is scaled by the scaling factor $\sqrt{d_k}$. Experiments done by the authors of the "Attention Is All You Need" paper show that the dot product alone could result in the vanishing gradient problem when there is a large variance between the values produced by the dot product.

$$S = QK^\top = \begin{bmatrix} \dfrac{q_1 \cdot k_1}{d_k} & \dfrac{q_1 \cdot k_2}{d_k} & \cdots & \dfrac{q_1 \cdot k_n}{d_k} \\[2mm] \dfrac{q_2 \cdot k_1}{d_k} & \dfrac{q_2 \cdot k_2}{d_k} & \cdots & \dfrac{q_2 \cdot k_n}{d_k} \\[2mm] \vdots & \vdots & \ddots & \vdots \\[2mm] \dfrac{q_m \cdot k_1}{d_k} & \dfrac{q_m \cdot k_2}{d_k} & \cdots & \dfrac{q_m \cdot k_n}{d_k} \end{bmatrix} \tag{7}$$

When such small values are fed to the softmax function, it returns small probabilities compared to the probabilities of the large values. As a result, their gradient becomes extremely small and leads to the vanishing gradient problem. Scaling the dot product by $\sqrt{d_k}$ ensures that the large values are scaled down to avoid large variances.

Let's explain this by a simple query vector of dimensionality $d_k = 4$. The scaling factor is:

$$\sqrt{d_k} = 2.0. \tag{8}$$

Given the query and key vectors:

$$Q_i = \begin{bmatrix} 2 \\ 4 \\ 1 \\ 3 \end{bmatrix}, \quad K_i = \begin{bmatrix} 1 \\ 3 \\ 2 \\ 4 \end{bmatrix} \tag{9}$$

Their dot product is:

$$Q_i \cdot K_i = (2)(1) + (4)(3) + (1)(2) + (3)(4) = 28. \tag{10}$$

The scaled dot product is:

$$\frac{Q_i \cdot K_i}{\sqrt{d_k}} = \frac{28}{2.0} = 14.0. \tag{11}$$

Without scaling, the dot product is 28, which can produce very large values during the softmax computation. With scaling, the dot product is reduced to 14.0 to make the softmax function outputs more stable. Table 2-4 shows the new matrix of dot products after applying the scaling factor d_k over each value.

Table 2-4. *Scaled dot product*

	W1	W2	W3	W4	W5	W6
W1	6.0	4.0	3.5	5.0	8.5	7.5
W2	4.5	4.5	1.0	4.5	8.0	4.5
W3	5.5	3.5	0.5	7.0	4.5	6.0
W4	5.0	3.0	2.5	0.5	6.0	1.5
W5	5.5	0.0	1.5	3.0	9.0	4.0
W6	0.5	1.5	2.5	1.0	3.0	2.5

After applying the softmax function with the scaling factor d_k, Table 2-5 shows the new scaled attention pattern.

Table 2-5. *Scaled attention pattern*

	W1	W2	W3	W4	W5	W6
W1	**0.3561**	0.2686	**0.4993**	0.1092	0.2976	**0.762**
W2	0.0795	**0.4428**	0.041	0.0662	0.1805	0.0379
W3	0.216	0.1629	0.0249	**0.8066**	0.0055	0.17
W4	0.131	0.0988	0.1837	0.0012	0.0244	0.0019
W5	0.216	0.0049	0.0676	0.0148	**0.4907**	0.023
W6	0.0015	0.022	0.1837	0.002	0.0012	0.0051

Softmax

Instead of having values ranging from $-\infty$ to ∞ or -1 to $+1$, we aim to bound them within the range 0 to $+1$. This transformation serves two purposes:

1. Scaling the numbers down to a consistent range.

2. Interpreting each value as a probability that indicates how strongly one word is related or similar to another.

The objective is to normalize each column of the matrix so that the sum of its elements equals 1.0, treating each column as a probability distribution. The token corresponding to the highest probability will be selected as the most likely candidate to alter the meaning of the current token.

A simple way to achieve this normalization is to divide each value by the sum of the entire column:

$$\text{Prob}(z_i) = \frac{z_i}{\sum_{j=1}^{n} z_j} \tag{12}$$

However, this approach may lead to issues, such as division by zero, if the sum of all values equals zero. This problem is addressed by applying the softmax function, which normalizes the scores into a probability distribution, ensuring that the sum of all values is always equal to 1.

$$\text{Softmax}(z_i) = \frac{e^{z_i}}{\sum_{j=1}^{n} e^{z_j}} \tag{13}$$

This resulting matrix, shown in Table 2-6, is called the attention pattern.

Table 2-6. *Attention pattern*

	W1	W2	W3	W4	W5	W6
W1	**0.5206**	0.2365	**0.7703**	0.0179	0.2443	**0.9495**
W2	0.0259	**0.6428**	0.0052	0.0066	0.0899	0.0024
W3	0.1915	0.087	0.0019	**0.9752**	0.0001	0.0473
W4	0.0705	0.032	0.1042	0.0	0.0016	0.0
W5	0.1915	0.0001	0.0141	0.0003	**0.6641**	0.0009
W6	0.0	0.0016	0.1042	0.0	0.0	0.0

Output

Remember that the V matrix size is $L \times d_v$ where each embedding vector is represented by a vector of length d_v. As a result, the output of self-attention is of such size. To accommodate situations where dv is not equal to d_{model}, this matrix has to be resized back to the original embedding size d_{model}. To do that, we have to use another weights matrix W_o of size $d_v \times d_{model}$ as in Figure 2-12.

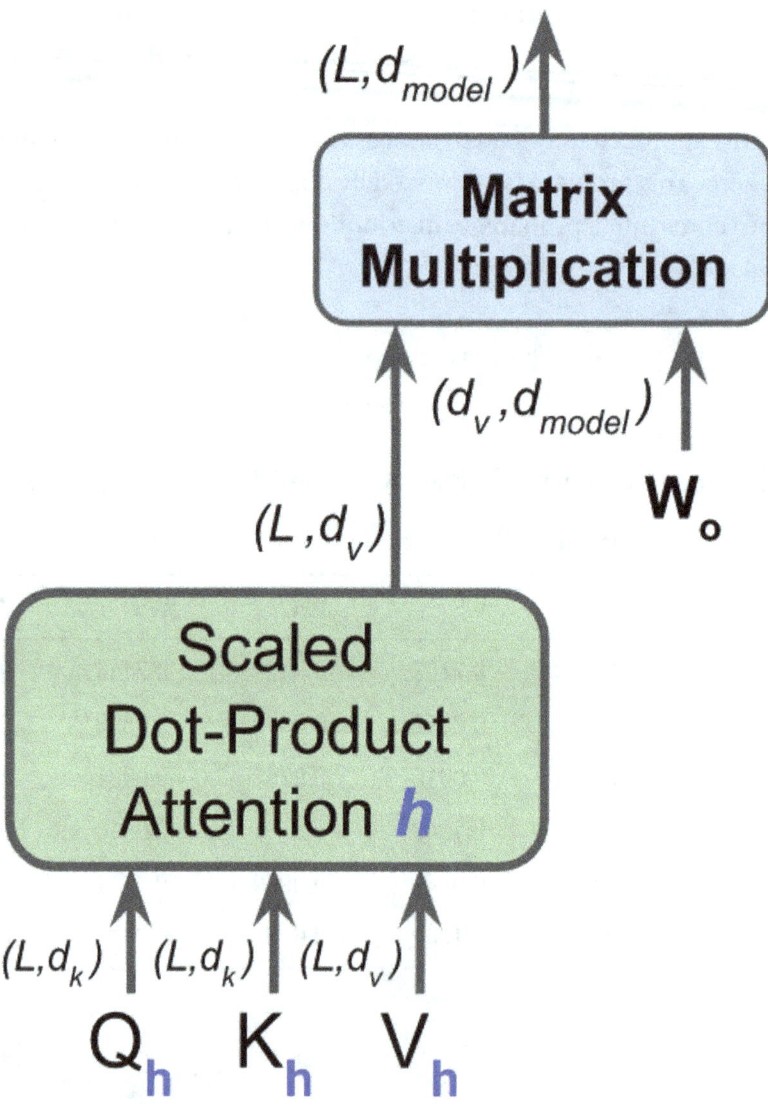

Figure 2-12. *The attention output is generated by multiplying the scaled dot-product attention result with the output weight matrix W_o*

The matrix W_o produces the outputs of self-attention to restore the original embedding of size $L \times d_{model}$.

$$AttentionOutput = \text{Attention}(Q,K,V) \times W_o \qquad (14)$$

If $d_{model} = d_v$, you might think that the trainable output weights matrix W_o is not needed. But it is still needed. One motivation to use it is because self-attention produces its output as a weighted sum between the softmax output and the value V matrix. This weighted sum is just a linear combination, and the matrix W_o is used to apply transformations over the features extracted by self-attention to learn more complex patterns. Figure 2-13 shows a simplified version of how the attention block works.

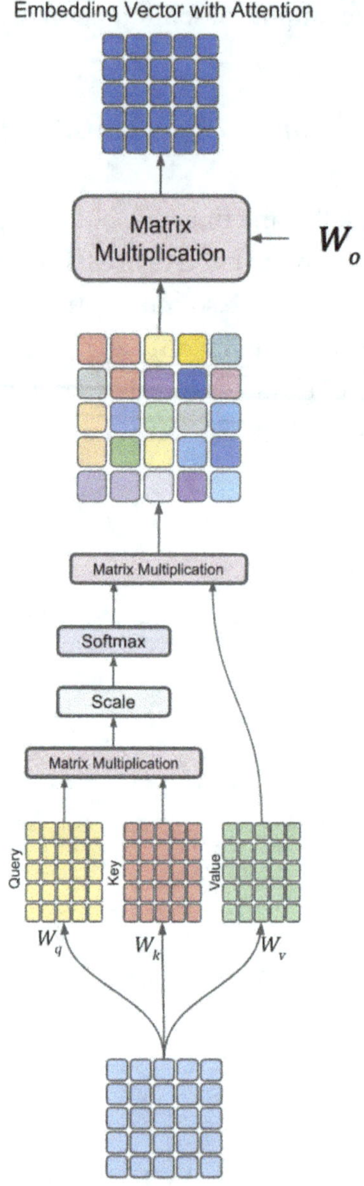

Figure 2-13. *Single-attention head. The input is first projected into query, key, and value matrices. The query and key are used to compute attention scores through a scaled dot product, followed by a softmax function to obtain weights. These weights are multiplied by the value matrix to produce the attention output, which is then transformed by the output weight matrix W_o to capture more complex feature interactions*

Later, the attention output is passed to the elementwise add layer, where each record in the attention output is added to its corresponding embedding vector E^i to translate the vectors into their d_{model}-dimensional space according to their context.

$$E^i_{new} = E^i_{old} + AttentionOutput^i \tag{15}$$

As an example, consider an embedding vector for the token *player*. Initially, this embedding vector does not indicate whether the player is associated with football, handball, computer games, or another context. By applying the attention mechanism, a vector is generated (e.g., *AttentionOutput^i*) that, when added to the embedding vector of the token *player*, enhances its representation to reflect, for example, that it denotes a handball player.

Number of Parameters

If the length of the query or key vector is d_k = 128 and the embedding vector length is d_{model} = 768, then the three weight matrices size is $d_{model} \times d_k$ = 768x128. This means we have 3 * 98, 304 = 294, 912 parameters just for the three matrices.

Additionally, the fourth weight matrix W_o is considered. The size of this matrix is $d_v \times d_{model}$ = 128x768 which adds another 98, 304 to the total parameters count. As a result, the total parameters count for a single head of attention is 4 * 98, 304 = 393, 216.

Coding Example

This section explains how to use the BaseBERT embedding model to tokenize and measure the attention of the following input text:

```
The quick brown fox jumped over the lazy dog while running as crazy.
```

The Hugging Face `transformers` library is employed to load the tokenizer. The uncased version is used, which ignores the case of the words, treating *Player* and *player* as identical.

```
import transformers

model_name = "bert-base-uncased"
tokenizer = transformers.BertTokenizer.from_pretrained(model_name)
```

The loaded tokenizer is applied to generate a list of tokens. By setting `return_tensors="pt"`, the tokenizer returns a PyTorch tensor containing the token IDs. Alternatively, the token IDs can be returned as a TensorFlow tensor by setting `return_tensors="tf"` or as a NumPy array by setting `return_tensors="np"`.

```
text = "The quick brown fox jumped over the lazy dog while running
as crazy."
inputs = tokenizer(text, return_tensors="pt")
```

The result is a dictionary containing three keys, each associated with a PyTorch tensor:

1. `input_ids`: Token IDs mapped according to the vocabulary used by the BaseBERT model.

2. `token_type_ids`: Sequence IDs. Since only a single sequence is used in this example, all values are zeros.

3. `attention_mask`: Indicates which tokens should be attended to and which should be ignored. This is particularly useful for distinguishing padding tokens from actual tokens, where 1 means attend and 0 means ignore. In our case, no padding is used.

```
{'input_ids': tensor([[101, 1996, 4248, 2829, 4419, 5598, 2058, 1996,13971,
3899, 2096, 2770, 2004, 4689, 1012, 102]]),
 'token_type_ids': tensor([[0, 0, 0, 0, 0, 0, 0, 0, 0, 0, 0, 0, 0, 0,
0, 0]]),
 'attention_mask': tensor([[1, 1, 1, 1, 1, 1, 1, 1, 1, 1, 1, 1, 1, 1,
1, 1]])}
```

The input sequence produces 16 tokens. The first token, `[CLS]` (Classifier Token), marks the beginning of the sequence, while the last token, `[SEP]` (Separator Token), indicates the end of the sequence.

```
['[CLS]','the','quick','brown','fox','jumped','over','the','lazy','dog',
'while','running','as','crazy','.','[SEP]']
```

Next, the pre-trained BaseBERT model is loaded. Setting output_attentions=True ensures that the attention weights are returned, which is essential for visualization. The token IDs and attention mask are then fed into the model to obtain the attention weights.

```python
import pytorch

model = transformers.BertModel.from_pretrained(model_name,
                                                output_attentions=True)
with torch.no_grad():
    outputs = model(input_ids=inputs['input_ids'],
                    attention_mask=inputs['attention_mask'])
attentions = outputs.attentions
```

The length of attentions is 12, corresponding to the number of attention blocks in the BaseBERT model. Each block applies multiple attention heads (which will be explained in the section "Multi-Head Attention"), with each head producing attention weights of shape [batch_size, num_heads, seq_length, seq_length].

Since there are 16 tokens in the input sequence, the output shape of each block is [1, 12, 16, 16], where:

- 1 represents the batch size,

- 12 is the number of attention heads,

- 16 is the number of tokens in the sequence.

To access the attention weights for a specific attention layer/block and head, you can index the attentions list accordingly.

```python
layer_num = 0
attention_head = 0
attention = attentions[layer_num][0, attention_head].cpu().numpy()
```

Figure 2-14 displays the heatmap of the attention pattern from the first attention head in the first attention block. The heatmap visualizes how much attention each token pays to other tokens in the sequence, helping to illustrate which tokens the model considers most relevant for each token. For example, the attention of the two tokens *fox* and *dog* is high, indicating that the model considers these tokens to be contextually significant to each other within the sentence.

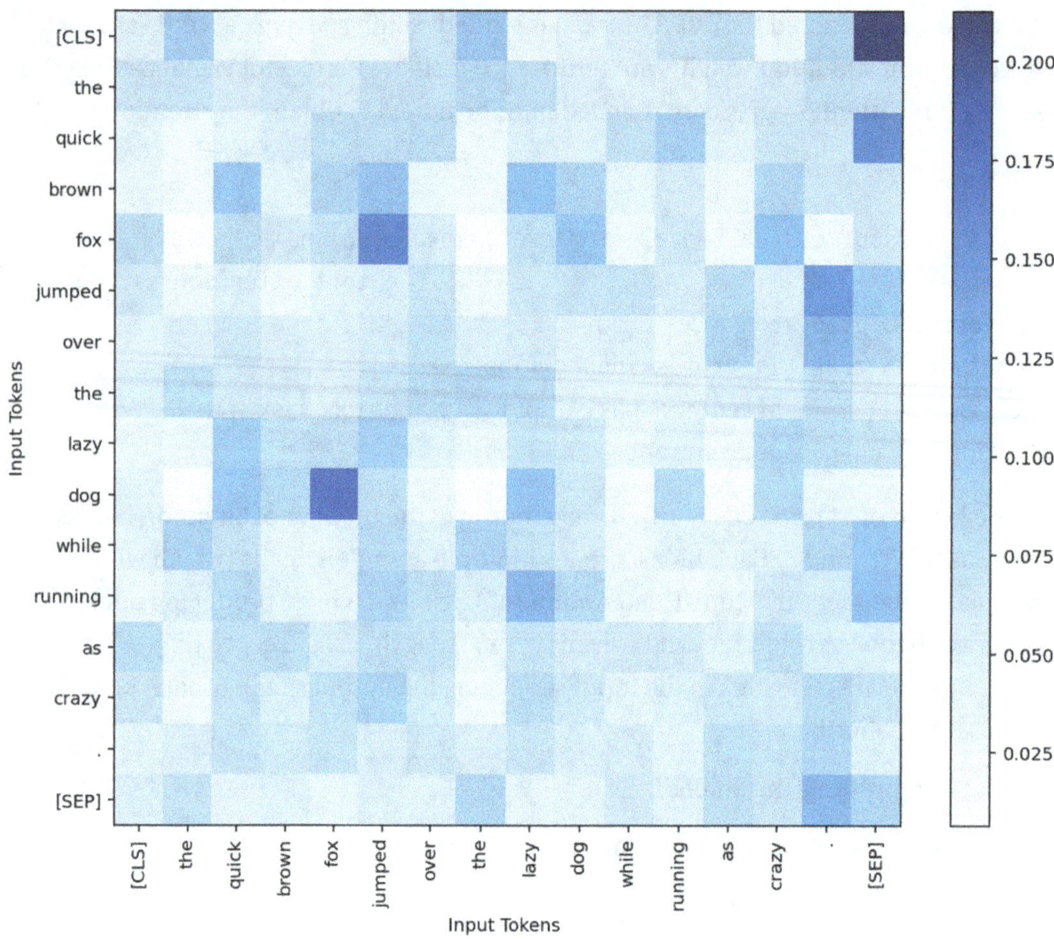

Figure 2-14. *Heatmap of the attention pattern from the first attention head in the first attention block. The heatmap shows how much each token attends to others in the sequence, highlighting the most contextually relevant relationships*

2.2.5 Multi-Head Attention

Single-head self-attention is applied by using three trainable weight matrices once over the embedding vector. However, a key limitation of single-head self-attention is that the learned matrices might ultimately focus on a single type of relationship between tokens. To address this, the authors of the *Attention Is All You Need* paper proposed multi-head attention (MHA), as in Figure 2-15. This approach applies multiple trainable matrices in parallel, with the objective of enabling each head to capture different aspects of the token relationships, thereby enhancing the model's capacity to represent diverse features.

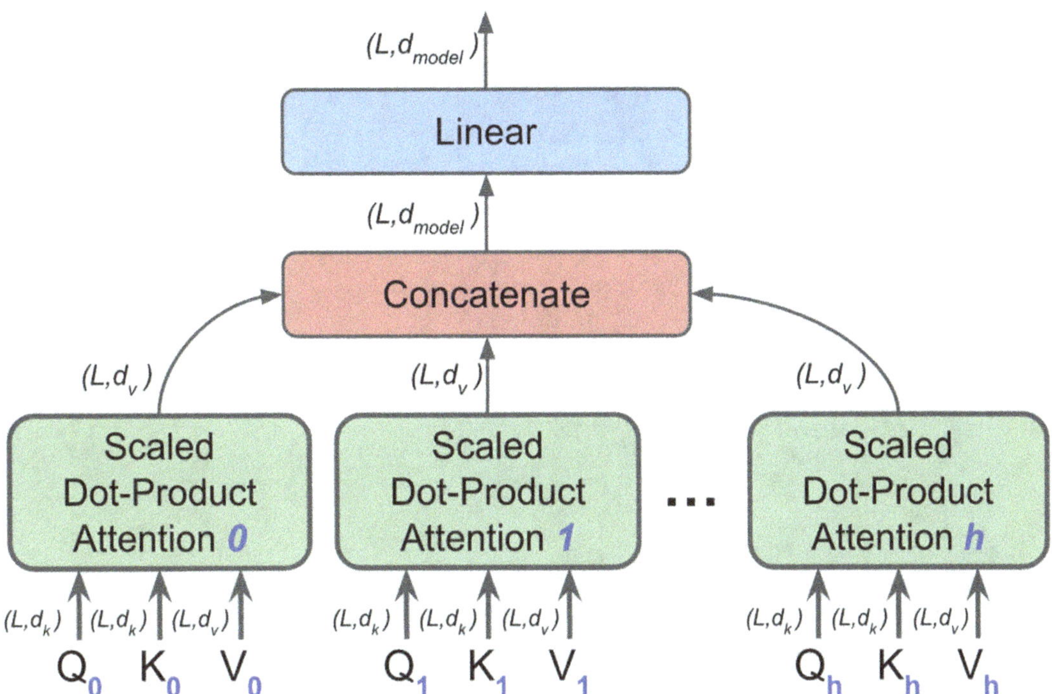

Figure 2-15. *Multi-head scaled dot-product attention. This mechanism runs multiple scaled dot-product attention operations in parallel, each with its own query, key, and value projections. The outputs from all heads are concatenated and transformed*

MHA can be thought of as applying the self-attention h times in parallel, where h is the number of heads. Instead of using separate full weight matrices W_q, W_k, and W_v for each head, they are split into h matrices, each one is used by a single head. This makes the computational cost of multi-head similar to the single-head self-attention since the number of parameters is still the same. Figure 2-16 shows a simplified diagram of the operations applied in the MHA block.

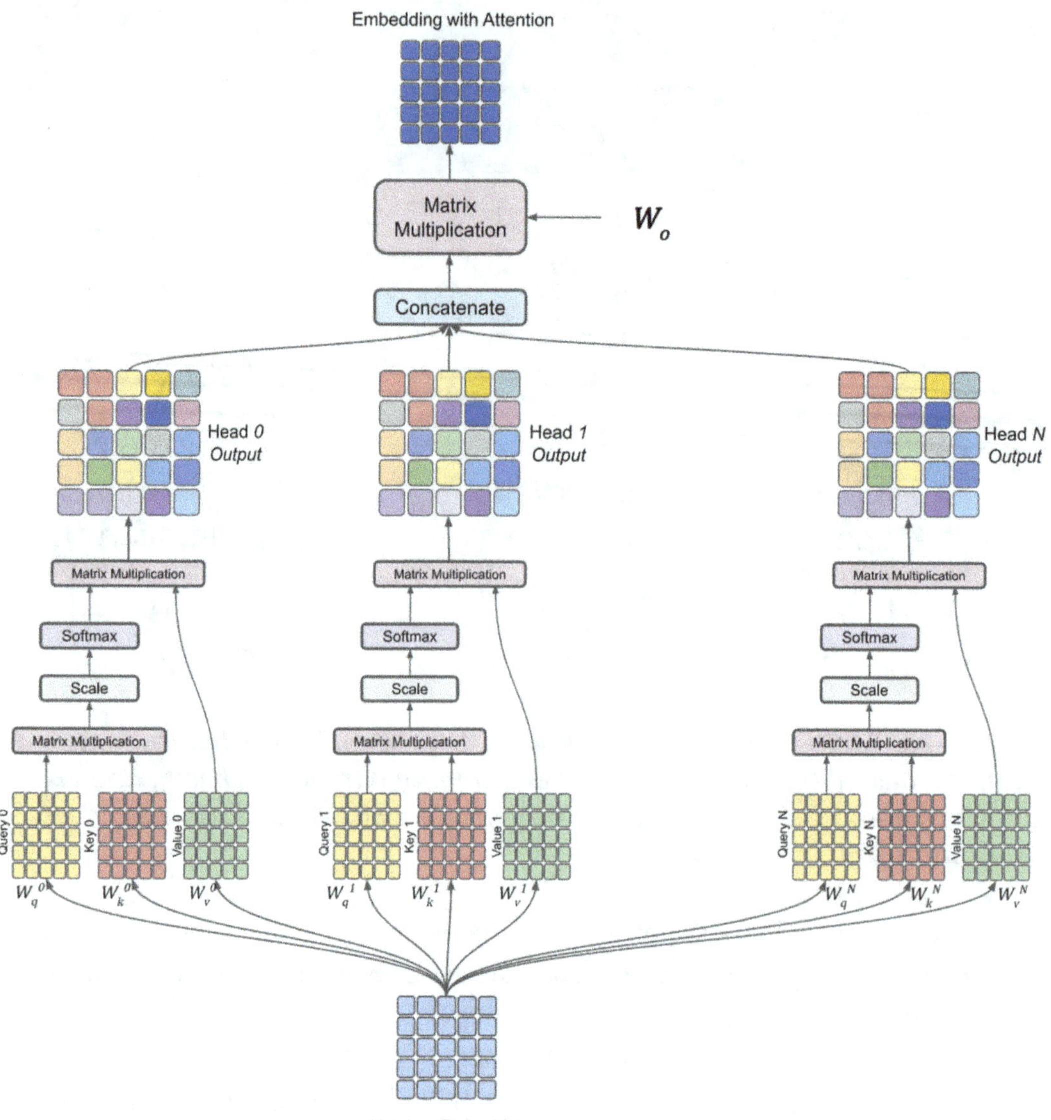

Figure 2-16. *Multi-head attention. The self-attention mechanism is applied h times in parallel, where h is the number of heads. The weight matrices W_q, W_k, and W_v are split into h smaller matrices, one for each head, keeping the total number of parameters and computational cost comparable to single-head attention*

If each full weight matrix is of size $d_{model} \times d_k$ where the embedding vector length is $d_{model} = 12,288$ and $h = 96$ (as in GPT-3), then the length of each vector in the query, key, and value is:

$$d_k = d_v = \frac{d_{model}}{h} = \frac{12,288}{96} = 128 \tag{16}$$

As a result, the size of a single weight matrix of one head is $d_{model} \times d_k = 12,288x128$. There are 96 small Query W_q^i, Key W_k^i, and Value W_v^i matrices, where i is the head index from 0 to $h - 1$. As a result, this creates h queries Q_i, keys K_i, and values V_i matrices, one for each head.

The scaled dot product attention is applied for each head. The result matrices of all the h heads are then concatenated to produce a single output of length hxd_k that is in turn equal to d_{model}. Remember that the size of each head output is $L \times d_k$, where L is the maximum number of tokens in the input sequence (i.e., 2048). After being concatenated, the result size is 2048×12288.

Think of each head as focusing on detecting a specific kind of relationship (e.g., syntactic, semantic, etc.) in the original embedding vector. The result embedding vector is simply the concatenation of all these attention heads results.

This result is then projected using the second value weights matrix W_o of size $d_{model} \times d_{model}$ (e.g., 12288×12288) to return the result of size $L \times d_{model}$ (e.g., 2048×12288). Recall that one objective of applying W_o is to transform the weighted sum produced by self-attention, enabling the model to capture more complex features. The result is then added to the original embedding vector to translate it in the d_{model}-dimensional space according to its context.

2.2.6 Add Layer

Up to this point, the raw embedding vectors are still used, which do not yet carry any information from their context. Once the self-attention is applied, it returns an output matrix of shape $L \times d_{model}$, where each row represents how each embedding vector should move in the d_{model}-embedding space according to its context.

Each row is a vector of the same length as the embedding vector, and their sum alters the direction of the embedding vector based on the context. Assume the input text contains the token *ball*. The generic embedding vector for this token represents a generic ball, similar to the blue ball depicted at the center of Figure 2-17.

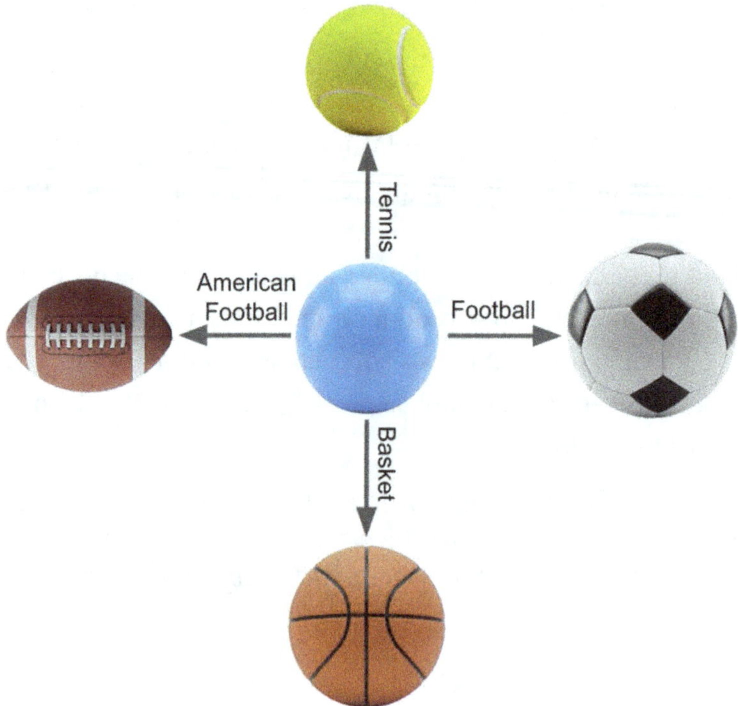

Figure 2-17. *Movement of the embedding vector for the token ball in the direction identified by the transformer's self-attention mechanism*

There are four possible directions the ball can move, each corresponding to the type of game in which the ball is used: football, tennis, American football, and basketball. Without context, the model cannot determine the appropriate direction for the ball (i.e., embedding vector). Based on the output of the self-attention mechanism, the row corresponding to the ball's embedding vector adjusts its direction according to the inferred game context. For example, if the input text is *tennis has a small ball*, the model infers that *ball* should move in the direction of the *tennis* game.

The transformer encoder feeds the self-attention output to an element-wise addition layer (refer to Figure 2-2). This layer accepts the following inputs:

1. Embedding matrix (after positioning encoding)

2. Self-attention output matrix

By adding these two matrices together (i.e., adding each embedding vector to its corresponding vector in the attention output), each generic embedding vector in the embedding matrix moves into the direction inferred from the context.

2.2.7 Normalization Layer

However, when the vectors are added together, their magnitudes might increase. Without control, this accumulation could lead to the exploding gradient problem, where the values become excessively large, destabilizing the network.

To mitigate this and restrict the range of values, the transformer applies layer normalization, which normalizes the input by centering it around zero and scaling it to have a standard deviation close to 1.0. The γ and β parameters are trainable to scale and shift the output values.

$$\mathrm{LayerNorm}(x) = \frac{x - \mu}{\sigma} \cdot \gamma + \beta \tag{17}$$

Where:

$$\mu = \frac{1}{d}\sum_{i=1}^{d} x_i$$
$$\sigma = \sqrt{\frac{1}{d}\sum_{i=1}^{d}(x_i - \mu)^2} \tag{18}$$

Compared to batch normalization, which calculates the mean and standard deviation based on a batch of samples, layer normalization works on each sample independently, and this is the motivation behind using it. It is common in NLP tasks to have a variable number of tokens in the input samples.

To illustrate this, assume there are two input samples with different numbers of tokens. The first sample contains only two tokens, while the second sample is misspelled, a common occurrence in NLP, resulting in five tokens. For simplicity, we exclude the special tokens CLS and SEP.

The tokenized samples are:

1.   Thank you: ['thank', 'you']

2.   Nounfortunately: ['noun', '##fort', '##una', '##tel', '##y']

Given that we only have two samples, then our current batch size is 2. In batch normalization, the statistics (mean and standard deviation) are calculated vertically across the i-th element of each token across all samples. For example, the first tokens of thank and noun from the two samples are grouped together, followed by you and ##fort. However, the remaining tokens in the second sample have no corresponding tokens in the first sample.

To handle this, there are two possible solutions:

1.   Exclude missing tokens from the first sample and calculate statistics only for the existing tokens in the second sample. This adds more overhead.

2.   Pad the first sample with special padding tokens so that both samples have an equal number of tokens.

Padding tokens are typically translated into zero values. Including many zeros in the normalization calculations can distort the statistics, leading to less accurate normalization and potentially harming model performance.

Layer normalization solves such issues by normalizing each token embedding vector independently across all its features, without considering other tokens or samples in the batch. Let's have a numerical example to make things clearer.

Given the input vector:

$$x = [2,4,6] \tag{19}$$

The mean μ is calculated as:

$$\mu = \frac{1}{3}(2+4+6) = \frac{12}{3} = 4 \tag{20}$$

The standard deviation σ is:

$$\sigma = \sqrt{\frac{1}{3}\left((2-4)^2 + (4-4)^2 + (6-4)^2\right)} = \sqrt{\frac{8}{3}} \approx 1.633 \qquad (21)$$

Next, the vector values are normalized using the mean and standard deviation:

$$\left[\frac{2-4}{1.633}, \frac{4-4}{1.633}, \frac{6-4}{1.633}\right] = \left[-1.224, 0, 1.224\right] \qquad (22)$$

Assuming the scale parameter $\gamma = 1$ and the shift parameter $\beta = 0$, the final normalized vector is:

$$\text{LayerNorm}(x) = \left[-1.224, 0, 1.224\right] \cdot 1 + 0 = \left[-1.224, 0, 1.224\right] \qquad (23)$$

The new vector has a mean of 0 and an approximate unit standard deviation, which helps stabilize the network and makes it resilient to the exploding gradients problem.

$$\mu = \frac{-1.224 + 0 + 1.224}{3} = 0 \qquad (24)$$

$$\sigma = \sqrt{\frac{(-1.224-0)^2 + (0-0)^2 + (1.224-0)^2}{3}} = \sqrt{\frac{1.498 + 0 + 1.498}{3}} \approx 0.999 \qquad (25)$$

2.2.8 Multi-Layer Perceptron (MLP)

According to the architecture diagram in Figure 2-2, each embedding vector undergoes the following transformations:

1. **Positional Encoding**: Captures the token position within the sequence.

2. **Self-Attention**: Utilizes scaled dot-product attention to determine the relevance of tokens to one another based on context.

3. **Add Layer**: Incorporates contextual information by translating the embedding vector into the d_{model}-dimensional space through element-wise addition.

4. **Normalization Layer**: Normalizes, scales, and shifts the embedding vector to stabilize the training process.

While the embedding vector at this stage possesses enhanced contextual awareness, which is the core innovation of the transformer model, there remains potential for further feature extraction and the recognition of more advanced patterns within the input sequence.

The primary limitation of self-attention is its reliance on linear operations, primarily matrix multiplications and element-wise additions. Linear transformations alone are insufficient for approximating complex non-linear functions.

Imagine data distribution as a highly complex structure. To approximate this structure with a mathematical function, the applied transformations must capture subtle distribution details. Linear operations alone cannot achieve this. Even a simple sine wave, as shown in Figure 2-18, cannot be accurately approximated without non-linear components. Consequently, non-linear transformations are essential for capturing deep patterns and uncovering hidden information within the embedding vector.

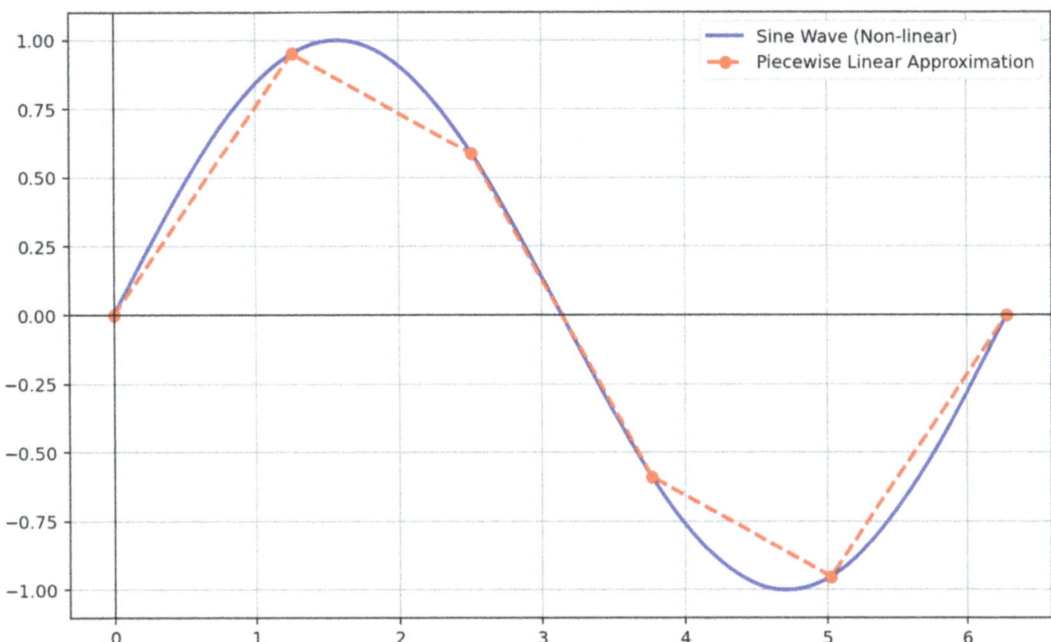

Figure 2-18. *Approximation of a sine wave. Linear operations alone cannot accurately model the wave, highlighting the need for non-linear transformations to capture complex patterns*

Multi-layer perceptrons (MLPs) are applied over the most recent version of the embedding vector (i.e., output of the normalization layer) as in Figure 2-19. MLP is a special kind of FFNN where it only uses dense layers and non-linear activation functions.

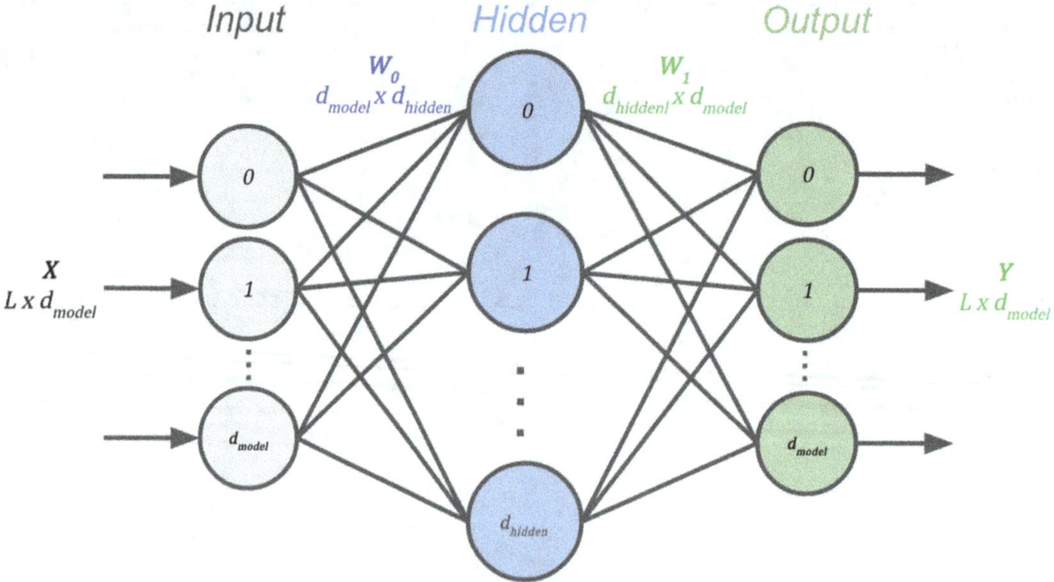

Figure 2-19. *Applying a multi-layer perceptron (MLP) to embedding vectors to transform and extract higher-level features*

The MLP used in transformers is relatively simple, consisting of a single hidden layer with the ReLU activation function. Its primary role is to apply non-linear transformations to the input, enabling the extraction of more complex patterns and features. The number of neurons in this hidden layer is typically x times larger than the embedding vector length, with a common practice of setting x to 4. For instance, if the embedding size is 512, the hidden layer would contain $512 \times 4 = 2048$ neurons. For GPT-3, the embedding size is 12, 288 and the number of hidden neurons is $12, 288 \times 4 = 49, 152$.

Following the MLP, an addition operation and layer normalization are applied, similar to those used after the self-attention mechanism, as illustrated in Figure 2-20. Finally, the resulting embedding matrix of size $L \times d_{model}$ is returned. In addition to the four previously listed aspects, each embedding vector now encodes richer information through non-linear transformations using the MLP and its ReLU activation function.

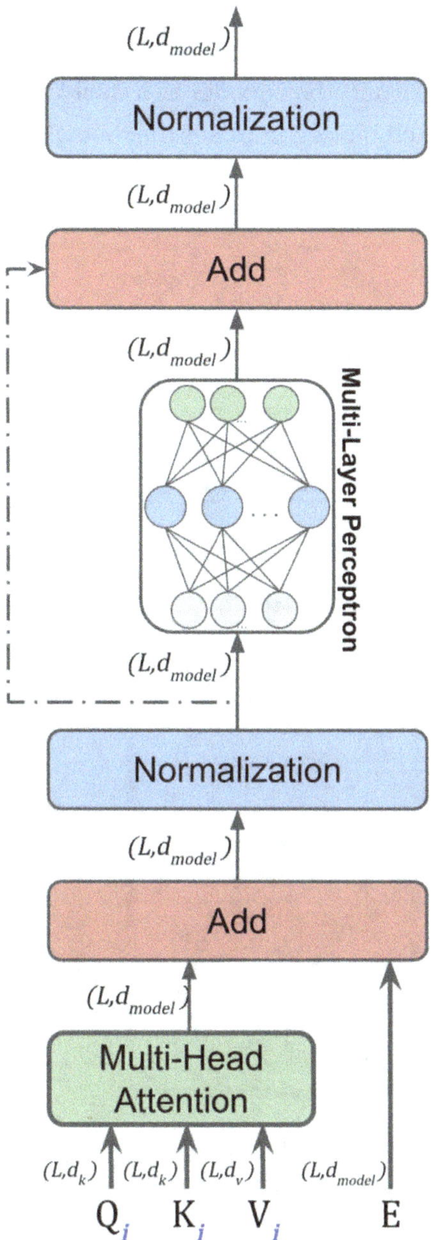

Figure 2-20. *Add and normalization layers applied after the multi-layer perceptron to stabilize training and maintain feature scale*

2.2.9 Layers Stack

The transformer model constructs the encoder as a stack of N identical blocks, as illustrated in Figure 2-21. Each block comprises the following layers:

1. Self-attention

2. Add and norm

3. Multi-layer perceptron (MLP)

4. Add and norm

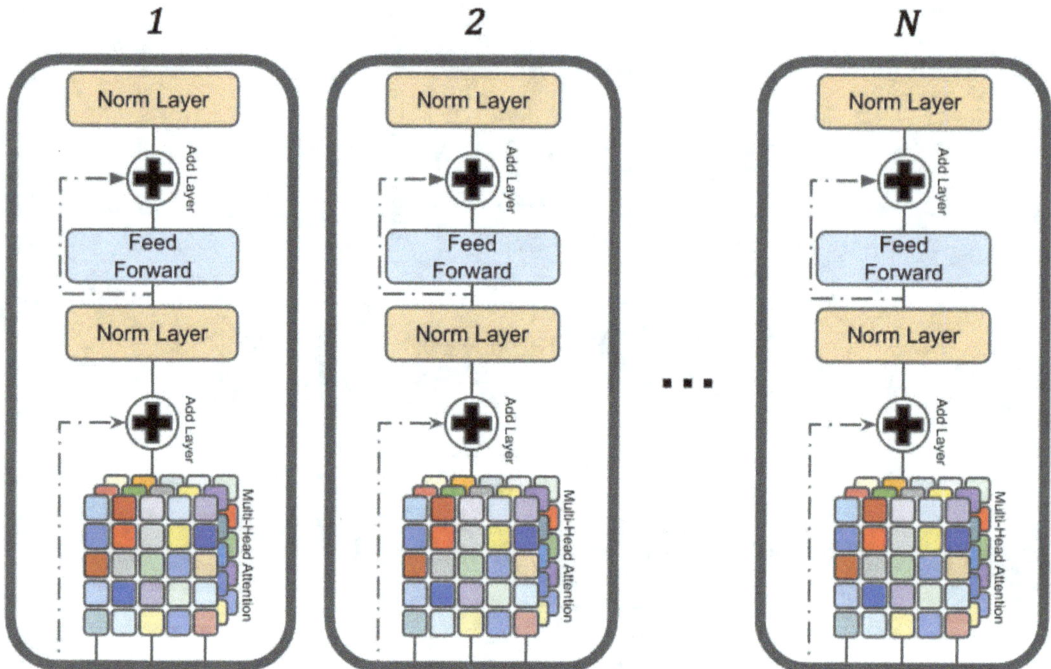

Figure 2-21. *Stack of encoder blocks, each containing self-attention, feed-forward, add, and normalization layers, arranged sequentially to progressively build richer contextual representations*

This stack enables the model to progressively extract more advanced features and enhance its contextual understanding. The repeated application of these blocks allows the encoder to capture both local and global features within the input sequence.

2.3 Key Value (KV) Cache

Before processing any tokens, the query (Q), key (K), and value (V) matrices in the attention mechanism are initially empty. For each input token fed to the transformer encoder, its embedding vector is multiplied by three separate weight matrices, one each for query, key, and value. These weight matrices are learnable parameters of the attention layer.

In practice, this multiplication is applied to the entire sequence of embeddings at once using matrix operations for efficiency. However, for simplicity, we will explain the process with a single token in mind.

In autoregressive generative models (e.g., GPT), the model generates one token at a time and computes attention over only the previously generated tokens. At the start of generation, there are no prior tokens, so a special start-of-sequence token (e.g., <s> or <|bos|>) is used to initiate the process. This token is encoded and passed to the decoder, which uses it to generate the first output token.

For each decoding step, the current token's embedding is transformed into a query, key, and value vector by multiplying it with the decoder's learnable weight matrices W_q, W_k, and W_v. That is:

$$q^i = x^i W_q, \quad k^i = x^i W_k, \quad v^i = x^i W_v \tag{26}$$

where x^i is the embedding of the i-th token in the sequence. After the first token is generated, the decoder's K, and V matrices will each contain a single row corresponding to that token. We are only interested in the key and value matrices, and the reason will be explained soon.

When a new token is generated, it is embedded and again passed through the decoder. At this point, the decoder needs to compute the query for the new token and attend to all previous tokens. This means it must have access to the key and value vectors of all previous tokens. If these key and value vectors are not cached, they must be recomputed at every decoding step, leading to redundant and inefficient computation.

For example:

- When the second token is generated, the decoder recomputes key and value vectors for both the first and second tokens.

- When the third token is generated, the decoder recomputes key and value vectors for the first, second, and third tokens.

So even if the key and value vectors for the tokens at the previous i steps are unchanged, they must be recomputed at the next step $i+1$. If we are at the fourth iteration as in Figure 2-22, this leads to the key and value vectors of earlier tokens being recomputed multiple times:

1. Four times for the first token

2. Three for the second

3. Twice for the third

4. Once for the fourth

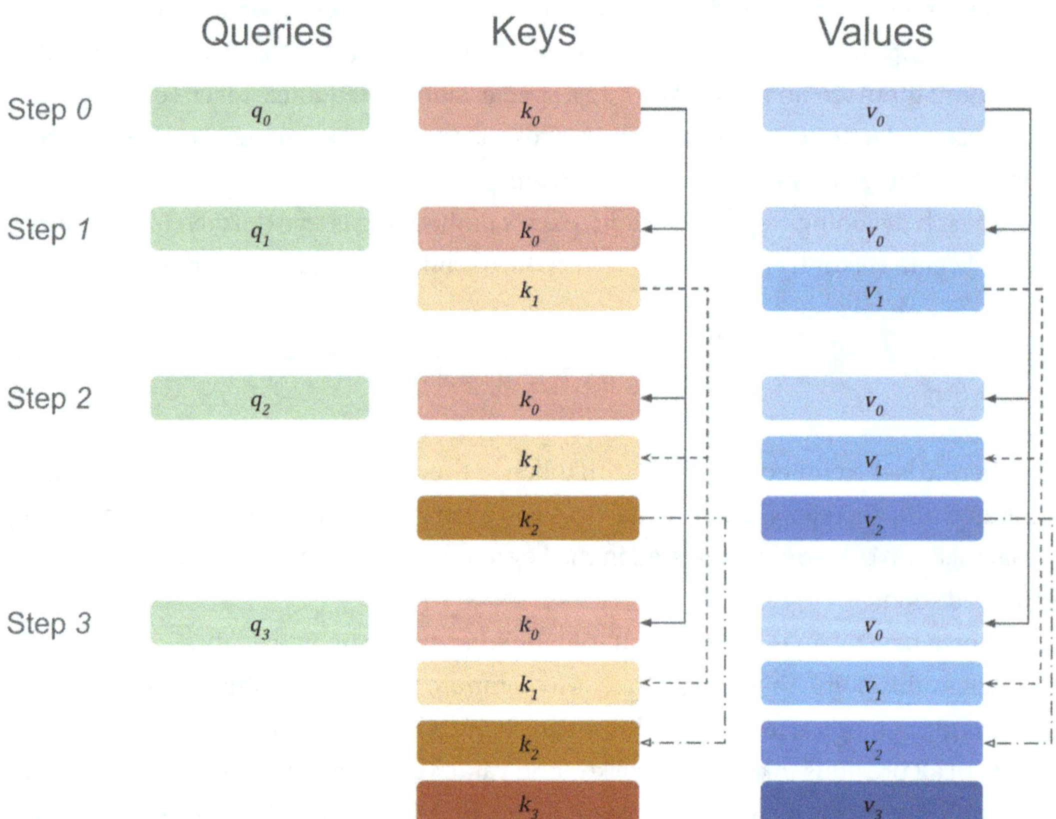

Figure 2-22. *Repeated use of key and value vectors for all previous tokens at each generation step in autoregressive generative models, enabling the model to maintain context and coherence throughout the sequence*

Such repeated computation is wasteful and becomes a major bottleneck during inference in long sequences. Think of it like you must read all the words you wrote before writing a new word. As the number of words you wrote increases, it becomes more tiresome and time-consuming. This could be saved by simply memorizing the words you wrote previously.

The query vector is calculated only once for each token because it is used immediately during that token decoding step. It depends only on the current token embedding and does not need to be reused.

To understand this intuitively, consider the analogy of a teacher preparing a question (query) to identify top students for a competition. The teacher needs to prepare the question only once. However, the question must be asked to each student (keys and values) to collect and evaluate their answers. The evaluation process determines which students are selected, similar to how attention scores determine which previous tokens the model should focus on when generating a new token.

In the same way, the query vector of the current token is multiplied with the key vectors of all previous tokens to compute attention scores. In practice, there are many query vectors used, and each searches for a property in the other tokens. The attention scores are then used to weight the value vectors and produce the final output for that token, following the scaled dot-product attention equation:

$$\text{Attention}(Q,K,V) = \text{softmax}\left(\frac{QK^T}{\sqrt{d_k}}\right)V \tag{27}$$

Since the key and value vectors only depend on the token embedding vector, which does not change, it is useless to recompute the key and value vectors for the same token each time. It is enough to calculate them only once and just reuse them. This is why modern transformer implementations use KV caching, which stores the key and value vectors after their initial computation so they can be reused without recomputation. This way we only need to calculate the key and value vectors once for the new token at each decoding step.

When an LLM receives a prompt and has not yet generated any tokens, it takes some time to produce the first token because it must compute the key and value vectors for every token in the prompt. At this stage, there is no KV cache available to reuse these vectors.

But KV caching comes at the cost of increasing memory requirements. Because there is a cache for each of the key and value vectors, *2* is used in the equation.

$$\text{KV Cache Size} = 2 \times L \times n \times h \times d_k \tag{28}$$

Where:

- n: number of tokens (sequence length)

- d_k: dimension of key and value vectors

- L: number of attention layers

- h: number of attention heads

Assuming a model with 128 heads, 64 layers, a dimension size of 128, 65,536 tokens, and floating-point 16 precision where each value is stored in 2 bytes, then the KV cache size is *256* GBs.

2.4 Encoder As a Feature Extractor

The encoder in a transformer generates embedding vectors for the tokens, which can be considered as feature vectors. They can be used to train ML models on tasks like classification.

Instead of using the individual feature vectors of the individual tokens, it is better to have a single vector representing the entire sequence. One way is to apply the pooling operation to create a single vector out of all the vectors. This is done by applying either max or average pooling to return a single vector. But their problem is ignoring the order of the tokens once pooling is applied. An alternative way is by using a token that represents the entire sequence. For example, BERT models have the *[CLS]* token representing the beginning of each input sequence. Its embedding vector encodes information from the entire sequence.

This section goes through a binary classification task using the Rotten Tomatoes dataset. This dataset contains nearly 10,000 samples, each labeled to indicate whether the tomatoes are rotten or not. The HuggingFace `datasets` library is used to load the dataset and the `transformers` library to tokenize the text and extract features using the DistilBERT model. Finally, a traditional machine learning model, specifically a Random Forest classifier, is trained to predict labels.

The Rotten Tomatoes dataset can be loaded using the `datasets` library.

```
import datasets

rotten_tomatoes = datasets.load_dataset("rotten_tomatoes")
```

The dataset is returned as a `DatasetDict` object, which contains three datasets: `train`, `test`, and `validation`. Each dataset has two keys:

1. `text`: A list of text samples.

2. `label`: A list of the labels.

```
DatasetDict({
    train: Dataset({
        features: ['text', 'label'],
        num_rows: 8530
    })
    validation: Dataset({
        features: ['text', 'label'],
        num_rows: 1066
    })
    test: Dataset({
        features: ['text', 'label'],
        num_rows: 1066
    })
})
```

Next, the tokenizer for the DistilBERT model is loaded to convert the text samples into tokenized sequences. A custom `tokenize()` function is created that truncates the samples to a specified maximum length and pads them to ensure all sequences have the same length. The tokenized sequences are returned as PyTorch tensors.

```
import transformers

tokenizer = transformers.AutoTokenizer.from_pretrained("distilbert/
distilbert-base-uncased")
```

```python
def tokenize(tokenizer, data, max_length=300):
    return tokenizer(data,
                     max_length=max_length,
                     truncation=True,
                     padding=True,
                     return_tensors="pt")

train_data_tokenized = tokenize(tokenizer=tokenizer,
                                data=rotten_tomatoes['train']['text'])
test_data_tokenized = tokenize(tokenizer=tokenizer,
                               data=rotten_tomatoes['test']['text'])
```

After tokenizing the data, the next step is to extract the features using the DistilBERT model. The feature array has a size of [batch, tokens, embedding_size], where each sequence has a number of feature vectors equal to the number of tokens. To solve this issue, the embedding vector of the first token of each sequence, [CLS], is considered the feature vector for that sequence. This reduces the array size to [batch, embedding_size].

```python
import torch

def extract_features(model, data_tokenized):
    with torch.no_grad():
        features = model(data_tokenized['input_ids'],
                         data_tokenized['attention_mask'])
        features = features.last_hidden_state[:, 0, :]
        return features

model = transformers.AutoModel.from_pretrained("distilbert/distilbert-base-uncased")
features_train = extract_features(model, train_data_tokenized)
features_test = extract_features(model, test_data_tokenized)
```

Using the extracted features, we can now train a traditional machine learning model. In this example, the Random Forest Classifier is used from the scikit-learn library.

```python
from sklearn.ensemble import RandomForestClassifier
from sklearn.metrics import accuracy_score
```

```
clf = RandomForestClassifier(max_depth=10, random_state=0)
clf.fit(features_train.numpy(), rotten_tomatoes['train']['label'])
y_pred = clf.predict(features_test.numpy())

score = accuracy_score(rotten_tomatoes['test']['label'], y_pred)
print(score)
```

2.5 Transformer Decoder

Generally, the transformer decoder consists of these steps:

1. Output embedding (shifted right)

2. Positional encoding

3. Stack of N layers

    1. Masked multi-head self-attention layer (triangle masking)

    2. Add and norm layer

    3. Multi-head cross-attention layer

    4. Add and norm layer

    5. Feed-forward neural network

    6. Add and norm layer

The decoder generates one token at a time in a sequential, layered fashion. The output of each decoder layer (i.e., block) serves as the input to the next layer, progressively refining the representation until all decoder layers have been processed. Each generated token, along with the most recent encoder output, is fed into the first decoder layer to predict the next token in the sequence.

When generating the first token, there is no previous token available from the decoder. To address this, a special starter token is used. This starter token is represented as an embedding vector, enabling the decoder to generate the first token solely based on the encoder's output.

The decoder architecture shares many components with the encoder, with a few key differences:

1. **Output Embedding**: The input sequence to the decoder is shifted to the right, ensuring that the decoder predicts the first token independently without being directly fed as an input.

2. **Masked Self-Attention**: The self-attention mechanism is masked to prevent the decoder from attending to future tokens, allowing it to consider only the previous tokens when predicting the next token.

3. **Cross-Attention**: This additional attention layer enables the decoder to measure the attention of the decoder's embedding against the encoder's output to generate the next token while taking the current input into consideration.

The next subsections will focus on explaining such differences.

2.5.1 Masked Scaled Dot-Product Attention

Transformers require vast amounts of data for effective training. To address this, their training strategy is designed to maximize the utilization of available training samples. For example, a single sentence like the one below is not treated as just one training sample for predicting the next word.

```
Handball players continue training daily ...
```

Instead, each token is treated as a separate training sample. By feeding the first token *Handball* to the model, each subsequent word is considered a training sample:

1. Handball ...

2. Handball players ...

3. Handball players continue ...

4. Handball players continue training ...

5. Handball players continue training daily ...

This approach allows a single training sample to generate multiple samples. For this process to work, each later token must not influence the earlier tokens. For example, when predicting the next word after *players*, the transformer model (specifically the decoder) should not receive any hint that the next word is *continue*. The model must deduce this on its own without using the attention pattern in Table 2-7.

Table 2-7. *Attention pattern for the handball players continue training daily sentence*

	Handball	**Players**	**Continue**	**Training**	**Daily**
Handball	0.7	0.05	0.12	0.07	0.2
Players	0.13	0.3	0.08	0.03	0.03
Continue	0.05	0.5	0.4	0.1	0.02
Training	0.07	0.2	0.3	0.6	0.25
Daily	0.05	0.05	0.1	0.1	0.4

By examining the attention pattern, each column represents the probabilities of all the words in the sentence. For the first column, the probability of the word *continue* is 0.5, which the model could interpret as a direct hint that the next word after *players* is *continue*. This situation could be interpreted as the model cheating by knowing the next word before predicting it. Such access prevents the model from learning to predict the next word independently.

The solution is to mask the probabilities in the attention pattern that correspond to future words for each token. This ensures that later words do not influence the decision-making process of earlier words. Table 2-8 highlights the probabilities that should be masked from the attention pattern, which simply corresponds to the lower half of the matrix. This is called masked scaled dot-product attention, which is used in the decoder block when predicting the next token. Refer to Figure 2-12. The only difference between it and its regular scaled dot-product attention is the masking step. Because the mask has a triangular shape, this kind of masking is referred to as triangle masking.

Table 2-8. *Highlighting the scores to be masked from the attention pattern*

	Handball	Players	Continue	Training	Daily
Handball	0.7	0.05	0.12	0.07	0.2
Players	**0.13**	0.3	0.08	0.03	0.03
Continue	**0.05**	**0.5**	0.4	0.1	0.02
Training	**0.07**	**0.2**	**0.3**	0.6	0.25
Daily	**0.05**	**0.05**	**0.1**	**0.1**	0.4

Such probabilities must be forced to zero. A simple approach is to replace these probabilities with zeros. Although this solves the issue, it introduces another problem where the sum of probabilities in each column will no longer equal 1.0. The correct method is to replace the probabilities of future tokens with $-\infty$ before applying the softmax function.

Table 2-9 shows the masked matrix before applying the softmax function, where the values represent the scaled dot products between the query and key vectors.

Table 2-9. *Replacing the next token probabilities by $-\infty$ before applying the softmax function*

	Handball	Players	Continue	Training	Daily
Handball	4.9	0.1	1.32	0.63	3
Players	$-\infty$	0.6	0.88	0.27	0.45
Continue	$-\infty$	$-\infty$	4.4	0.9	0.3
Training	$-\infty$	$-\infty$	$-\infty$	5.4	3.75
Daily	$-\infty$	$-\infty$	$-\infty$	$-\infty$	6

After applying the softmax function, Table 2-10 shows the new attention pattern, where all $-\infty$ values are replaced by zeros. This ensures the two conditions are met:

1. The masked probabilities do not contribute to the final attention scores.

2. The sum of probabilities is 1.0.

Table 2-10. *The attention pattern after masking the next token probabilities*

	Handball	Players	Continue	Training	Daily
Handball	1.0	0.3775	0.0427	0.0083	0.2222
Players	**0.0**	0.6225	0.0275	0.0058	0.0333
Continue	**0.0**	**0.0**	0.9298	0.0108	0.0222
Training	**0.0**	**0.0**	**0.0**	0.9751	0.2778
Daily	**0.0**	**0.0**	**0.0**	**0.0**	0.4444

Masking the probabilities of the next tokens preserves the auto-regressive nature of the model, ensuring that new tokens are generated solely based on the previously generated tokens. This technique prevents the model from "cheating" by having access to future tokens before making its predictions.

Given the masked attention pattern, the model determines which tokens are relevant to others. Based on this information, the embedding vectors of the tokens are adjusted to reflect their contextual similarity. To illustrate this, the token *players* has two probabilities that are not equal to zero.

	Players
Handball	0.3775
Players	0.6225

This means that the token *players* is represented as:

$$0.3775(handball) + 0.6225(players) \tag{29}$$

The token *players* is represented by taking 37.75% of the embedding vector of the token *handball* and 62.25% of the token *players*. If the old embedding vectors of the 2 tokens are:

$$\begin{aligned} handball &: [1,1] \\ playrs &: [5,5] \end{aligned} \tag{30}$$

Then the new embedding vector of the token *players* is:

$$
\begin{aligned}
0.3775(&handball)+0.6225(players)\\
&=0.3775*[1,1]+0.6225*[5,5]\\
&=[0.3775,0.3775]+[3.1125,3.1125]\\
&=[3.49,3.49]
\end{aligned}
\tag{31}
$$

The new embedding vector for the token *players* becomes:

$$
[3.49,3.49]
\tag{32}
$$

This transformation brings the token *players* closer to *handball* in the embedding space while distancing it from unrelated tokens like *football, computer,* or *tennis*. This demonstrates how the model uses context to update the embedding vector, making it reflect that *players* is more closely associated with *handball* than with other games.

2.5.2 Cross-Attention

Self-attention is applied to a single sequence to measure how each token attends or relates to other tokens within the same sequence. This mechanism is particularly useful in applications where the model processes only one sequence, such as predicting the next word in a sentence.

However, in applications like machine translation, where the model is trained with two different sequences (one fed into the encoder and the other into the decoder) representing the same text in two different languages (e.g., Arabic and English), self-attention alone is insufficient. Instead, cross-attention is applied to allow the decoder to attend to the encoder's output, enabling the model to map information between the two sequences.

The primary difference between self-attention and cross-attention lies in how the Q, K, and V matrices are computed, as in Figure 2-23. In self-attention, all three matrices are derived from the same sequence, enabling the model to capture dependencies within that sequence. In cross-attention, the Q matrix is extracted from one sequence (typically the decoder's input), while the K and V matrices are derived from another sequence (typically the encoder's output).

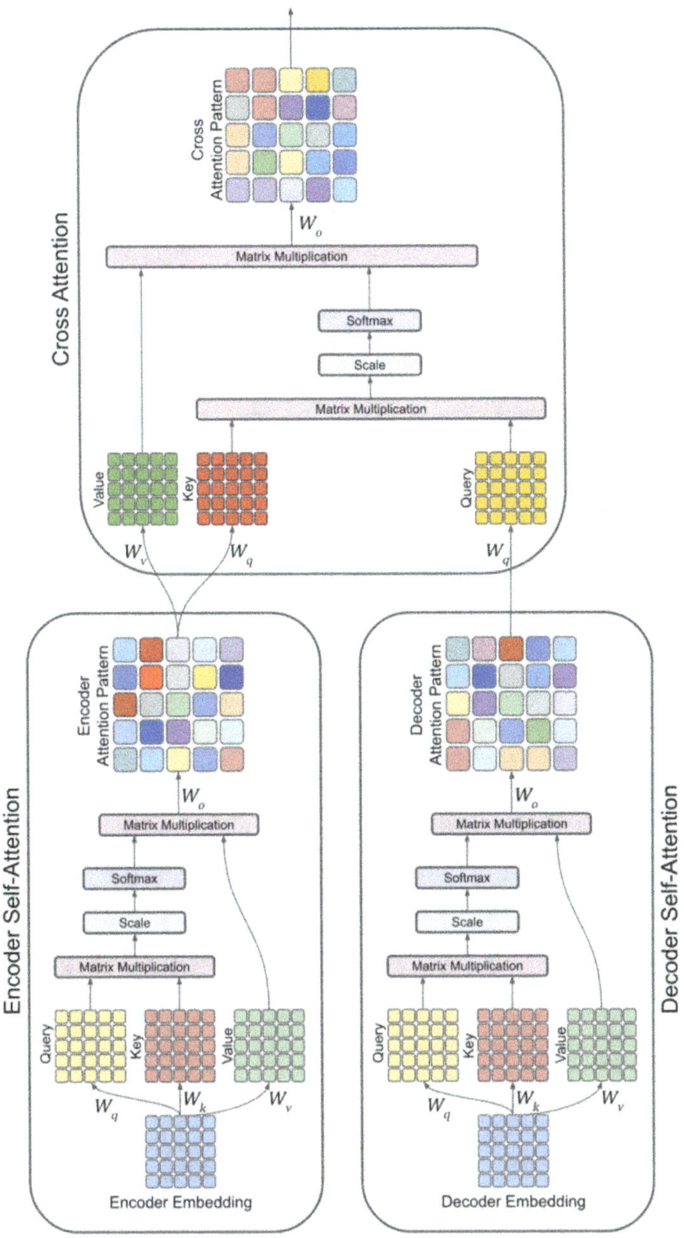

Figure 2-23. *Cross-attention. Unlike self-attention, where the Q, K, and V matrices are computed from the same sequence, cross-attention derives Q from one sequence (usually the decoder input) and K and V from another (usually the encoder output)*

The encoder's recent embedding matrix $E_{encoder}$ is multiplied by $W_k^{encoder}$ and $W_v^{encoder}$ to produce the matrices K and V, respectively. The decoder's recent embedding matrix $E_{decoder}$ is multiplied by $W_q^{decoder}$ to produce Q.

$$
\begin{aligned}
K &= E_{encoder} W_k^{encoder} \\
V &= E_{encoder} W_v^{encoder} \\
Q &= E_{decoder} W_q^{decoder}
\end{aligned}
\tag{33}
$$

During cross-attention, the encoder applies self-attention to the input sequence in one language, producing contextual embeddings. Simultaneously, the decoder applies self-attention to the target sequence. The outputs of both self-attention layers are then passed to the cross-attention layer, where the decoder attends to the encoder's output.

Once the Q, K, and V matrices are extracted, the attention calculation follows the same process in both self- and cross-attention, with the exception that masking is typically not applied in cross-attention.

Coding Example

This section demonstrates how to visualize the cross-attention matrix of Facebook BART (Bidirectional and Auto-Regressive Transformers) using the Hugging Face `transformers` library. BART is a powerful model for text generation and summarization. The specific model used in this example is `facebook/bart-large-cnn`.

To load the tokenizer, use the `AutoTokenizer` class. This class automatically selects the appropriate tokenizer based on the model's name.

```python
import transformers

model_name = "facebook/bart-large-cnn"
tokenizer = transformers.AutoTokenizer.from_pretrained(model_name)
```

If you know the exact tokenizer used by the model, you can load it directly using the model-specific class:

```python
tokenizer = transformers.BartTokenizer.from_pretrained(model_name)
```

Next, tokenize both the input and target text. The tokenizer returns PyTorch tensors representing the tokenized sequences. The sequence of tokens IDs is converted into strings using the `convert_ids_to_tokens()` method. Since there is only one batch, the token ID tensors have a shape of [1, sequence_length]. To simplify the tensor, the `squeeze()` method is applied to remove the batch dimension.

```python
input_text = "I want coffee not juice"
target_text = "I prefer coffee"

inputs = tokenizer(input_text, return_tensors="pt")
targets = tokenizer(target_text, return_tensors="pt")

tokens_inputs = tokenizer.convert_ids_to_tokens(inputs["input_ids"].
squeeze())
tokens_targets = tokenizer.convert_ids_to_tokens(targets["input_ids"].
squeeze())
```

The tokenized sequences are represented as both token IDs and their corresponding tokens. The Ġ character is a special marker used by the BPE tokenizer to indicate that the token is preceded by a space. Tokens at the start of the sequence, like I, do not include this marker.

```python
tensor([[0, 100, 236, 3895, 45, 10580, 2]])
['<s>', 'I', 'Ġwant', 'Ġcoffee', 'Ġnot', 'Ġjuice', '</s>']

tensor([[0, 100, 6573, 3895, 2]])
['<s>', 'I', 'Ġprefer', 'Ġcoffee', '</s>']
```

Next, load the BART model and feed the input and target sequences. Set `output_attentions=True` to ask the model to return the attention matrices.

```python
model = transformers.AutoModelForSeq2SeqLM.from_pretrained(model_name)
with torch.no_grad():
    outputs = model(**inputs, decoder_input_ids=targets["input_ids"],
    output_attentions=True)
```

The model's output object contains the following attention attributes:

1. `encoder_attentions`: A tuple of 12 elements representing encoder blocks, each with shape [1, 16, 7, 7] (1 batch, 16 encoder self-attention heads, 7 input tokens).

2. `decoder_attentions`: A tuple of 12 elements representing decoder blocks, each with shape [1, 16, 5, 5] (1 batch, 16 decoder self-attention heads, 5 target tokens).

3. `cross_attentions`: A tuple of 12 elements representing decoder blocks, each with shape [1, 16, 5, 7] (1 batch, 16 decoder cross-attention heads, 5 target tokens attending to 7 input tokens).

To visualize the cross-attention weights, select the attention matrix from a specific head (index 9) in a specific block (index 6):

```
cross_attention_weights = outputs.cross_attentions[6][0, 9].cpu().numpy()
```

The resulting heatmap, shown in Figure 2-24, highlights how the model aligns the input and target tokens. Notably, the token `prefer` in the target sequence attends to `want` in the input sequence, capturing the semantic relationship between the two.

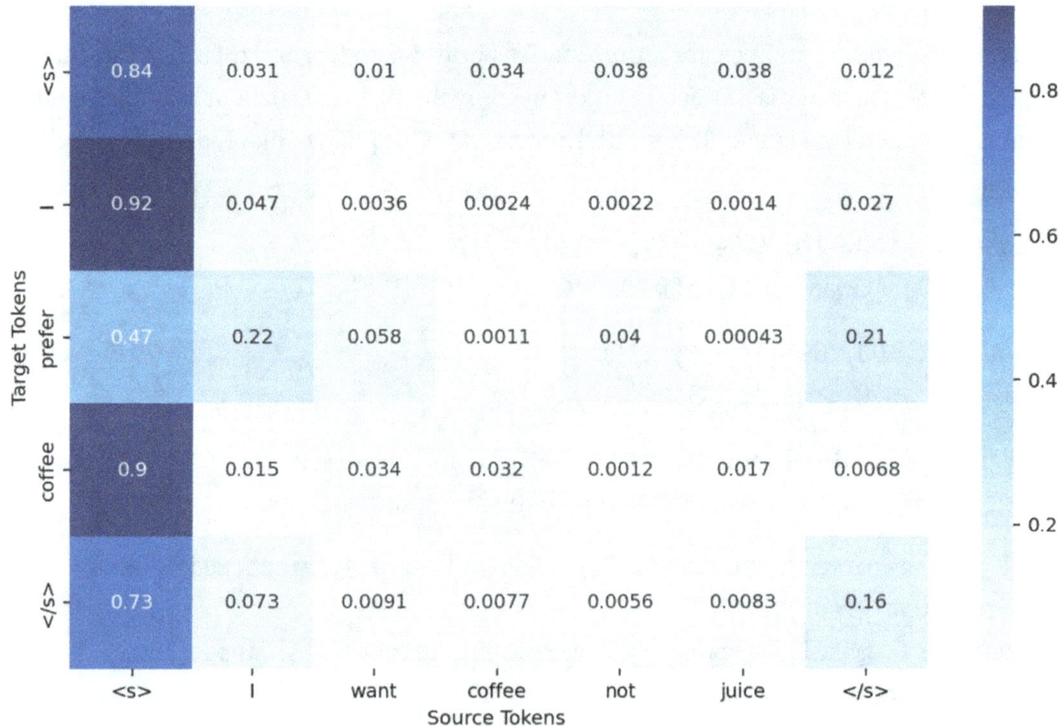

Figure 2-24. Cross-attention heatmap showing how the model aligns input and target tokens. For example, the token prefer in the target sequence attends to want in the input sequence, capturing their semantic relationship

2.5.3 Linear and Softmax Layers

At the end of the transformer model architecture, as in Figure 2-2, two layers are used to generate the model's final predictions:

1. Linear layer

2. Softmax layer

These two layers can be collectively regarded as a single linear (i.e., dense) layer with a softmax activation function. Their primary purpose is to utilize the decoder's output to predict the next token from the model's vocabulary, as illustrated in Figure 2-25.

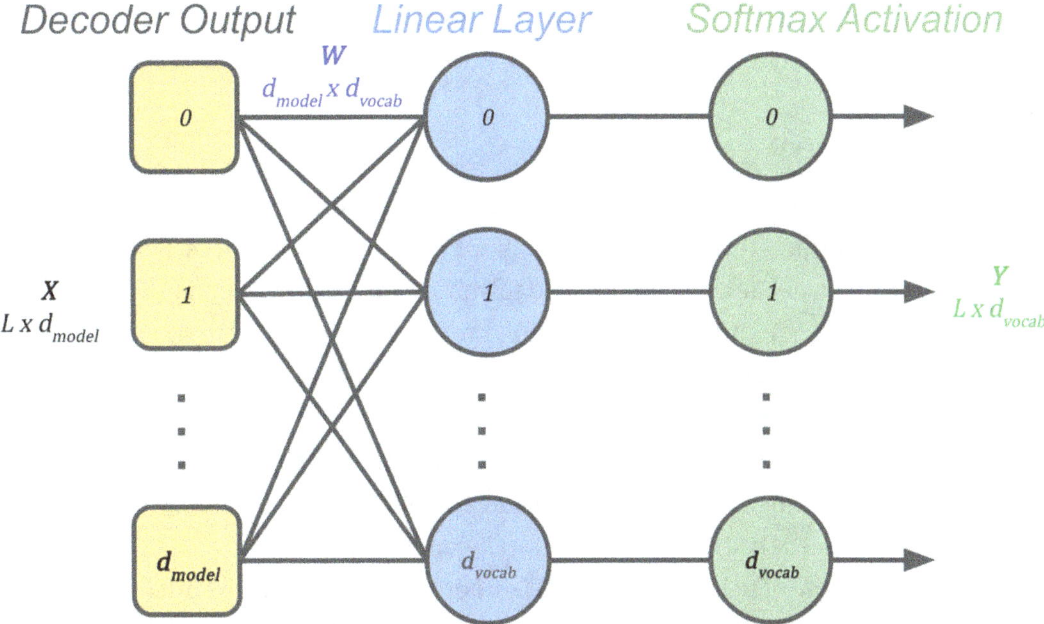

Figure 2-25. *Linear and softmax layers applied to the decoder's output to convert the final hidden states into probability distributions over the vocabulary for token prediction*

The output of the decoder's final layer (typically the normalization layer) is a vector of length equal to the embedding size, denoted by d_{model}. However, to predict the next token, this d_{model}-dimensional vector must be transformed into another vector whose length corresponds to the vocabulary size, d_{vocab}. This transformation is performed by the linear layer, which applies a learned weight matrix of size $d_{model} \times d_{vocab}$.

Subsequently, the softmax activation function is applied to the linear layer's output, converting the raw logits into a probability distribution over the entire vocabulary. The token with the highest probability is selected as the model's prediction for the next token in the sequence.

For example, if the vocabulary only has 3 tokens and the softmax probabilities are:

$$[0.1, 0.05, 0.895] \tag{34}$$

Then the selected word is football given that the vocabulary is:

```
| Token    | ID |
|----------|----|
| soccer   | 1  |
| good     | 2  |
| football | 3  |
```

The greedy decoding method, which always selects the token with the highest probability, is not always the optimal approach for generating sequences. Selecting the most probable token at each step might lead to suboptimal sentence construction. A more effective approach is to select tokens that produce the best overall sequence, even if they do not have the highest probability at each individual step. For this purpose, the beam search algorithm is often employed.

Beam search explores multiple possible sequences at each step and retains the most promising ones, resulting in higher-quality text generation. At each generation step, the search algorithm uses K beams to keep the top K generated tokens.

This is an example of using the Google T5 model for generating text out of the prompt *Translate from English to French: I like football.* The generate() method has the parameter num_beams, which is the number of beams in the beam search algorithm.

```python
import transformers

model_name = "t5-small"
tokenizer = transformers.AutoTokenizer.from_pretrained(model_name)
model = transformers.AutoModelForSeq2SeqLM.from_pretrained(model_name)

input_text = "Translate from English to French: I like football!"
inputs = tokenizer(input_text, return_tensors="pt")

num_beams = 2
```

```
outputs = model.generate(**inputs,
                         max_length=50,
                         num_beams=num_beams,
                         num_return_sequences=num_beams)

for beam in range(len(outputs)):
    generated_text = tokenizer.decode(outputs[beam],
                                      skip_special_tokens=True)
    print(f"Beam {beam}: {generated_text}")
```

To explore the branches of the two beams, the num_return_sequences parameter is set equal to the number of beams to return all the two different branches sequences.

```
Beam 0: J'aime le football!
Beam 1: Je veux le football!
```

2.5.4 Sampling

Additionally, some models use a decoding method called sampling decoding. This method controls how the model samples tokens during generation by using the following parameters:

1. Temperature

2. Top-K

3. Top-P

For any of these sampling methods to work, the do_sample parameter must be set to True in the generate() method.

In temperature sampling, the temperature parameter is introduced into the softmax function to control the randomness of token selection. The temperature range varies, but it could be 0 to 5 or 0 to 1. It controls the balance between randomness and creativity as shown in the following equation. T is the temperature that scales the inputs to the softmax function before applying the exponential function.

$$\text{Softmax}(x_i) = \frac{\exp\left(\dfrac{x_i}{T}\right)}{\sum_{j=1}^{K}\exp\left(\dfrac{x_j}{T}\right)} \tag{35}$$

Lower values of the temperature parameter T (i.e., close to 0) make the softmax output more deterministic, as the same input consistently yields the same output. This is because dividing each input logit x_i by a small value boosts its magnitude, resulting in a larger value after applying the exponential function. Consequently, the corresponding probability becomes significantly higher, leading to a sharper and more peaked distribution.

In contrast, higher values of T increase the randomness of the output by assigning more similar probabilities to different tokens, including those with lower original logits. Dividing x_i by a large value reduces its magnitude, and the exponential function yields smaller, more uniform outputs. This results in a smoother probability distribution and introduces diversity in the generated outputs. A very large value of T causes the output to be almost random because the tokens almost have the same probability.

This randomness can help generate more creative and varied text. For example, with a higher temperature, the model might select the word *soccer* instead of the more commonly used *football* in a given context. In the *transformers* library, this is supported using the `temperature` parameter in the `generate()` method. It ranges from 0 to ∞. The default value is 1.0.

```
output = model.generate(**inputs,
                        ...,
                        do_sample=True,
                        temperature=0.9)
```

Here are some of the different outputs returned:

- J'aime le football!

- Je veux me plier au football!

- Je me soyez attentif au football!

- Je veux parler du football!

- Je suis un joli footballeur!

For Top-K and Top-P sampling methods, the tokens are sorted in descending order according to their probabilities.

Token	Probability
play	0.6
game	0.2
computer	0.1
match	0.05
watch	0.05

In Top-K sampling, the model selects the next token from the K most probable tokens. If $K = 3$, the model restricts its choices to the three highest-probability tokens *play*, *game*, or *computer*. In the *transformers* library, this can be adjusted using the top_k parameter.

```
output = model.generate(**inputs,
                        ...,
                        do_sample=True,
                        top_k=10)
```

In Top-P sampling, the model selects the next token from the smallest set of tokens whose cumulative probability exceeds a threshold P. For instance, when top_p=0.7, the model narrows its choices to the tokens *play* and *game*, which together represent at least 70% of the total probability mass. Setting top_p=1.0 has no filtering effect, as it considers the entire vocabulary, making the sampling process equivalent to standard random sampling based on the model's probability distribution.

```
output = model.generate(**inputs,
                        ...,
                        do_sample=True,
                        top_p=0.7)
```

This marks the conclusion of our explanation of the traditional Transformer model architecture. Throughout the discussion, we have explored the key components and processes that define how a Transformer works.

We began with tokenization, the process of breaking down raw input text into smaller, meaningful units, called tokens, that can be processed by the model. These tokens were then transformed into dense numerical representations through an embedding layer, allowing the model to work in a high-dimensional space that captures semantic relationships.

Next, we examined the encoding phase, where each token representation is passed through multiple layers composed of self-attention mechanisms and feed-forward neural networks. The attention mechanism enables the model to dynamically weigh the importance of different tokens relative to each other, capturing contextual relationships regardless of their position in the sequence.

Finally, we covered how the encoded representations are used to generate output tokens through a decoder in sequence-to-sequence models, or directly from the encoder in tasks like classification or masked token prediction.

The architecture we explored forms the foundation of many modern language models, including BERT and GPT. A solid understanding of these core concepts is crucial before progressing to more advanced topics such as fine-tuning, transfer learning, and architectural variations such as multi-query attention, rotary positional embeddings, and mixtures of experts.

It also lays out the foundation for understanding the internal designs of models developed by major organizations, such as LLaMA, DeepSeek, and others.

CHAPTER 3

Advanced Transformer Architectures

Since the landmark release of the paper titled "Attention Is All You Need" in 2017 authored by a team of eight researchers, where six in total represented Google Brain and Google Research, the transformer model architecture has fundamentally redefined the landscape of artificial intelligence. This paper is considered one of the most recent influential works in the history of machine learning because it introduced the self-attention mechanism, allowing models to process entire sequences of data in parallel rather than sequentially. This shift solved the primary bottleneck of previous architectures like RNNs and LSTMs, which struggled with long-range dependencies and slow training speeds.

The transformer model architecture has undergone numerous enhancements aimed at improving its efficiency, scalability, and adaptability across tasks. These improvements span multiple areas, including optimized attention mechanisms to reduce computational cost, integration of richer positional encoding schemes, and innovations in training strategies to improve convergence and generalization.

This chapter explores some of the most notable improvements over the base architecture, highlighting both the motivations behind them and their practical impact on real-world applications.

3.1 Rotary Positional Embedding (RoPE)

Rotary positional embedding (RoPE) is a positional encoding technique that integrates position information directly into the token embedding vectors, rather than relying on a separately generated positional vector, as in traditional sinusoidal embeddings. It was introduced in the paper titled "RoFormer: Enhanced Transformer with Rotary Position Embedding" in 2021 by researchers in the AI startup called Zhuiyi Technology.

© Ahmed Fawzy Gad 2026
A. F. Gad, *Transformers and Large Language Models*, https://doi.org/10.1007/979-8-8688-2785-3_3

In traditional sinusoidal positional encoding, each token is assigned a unique absolute positional vector generated using sine and cosine functions. These vectors follow specific mathematical patterns that allow the neural network to infer the position of each token within the input sequence.

Because these vectors are based solely on the token's position in the sequence, this method is referred to as absolute positional encoding. The positional vector for a token is fixed and does not depend on the positions of surrounding tokens.

In this method, the positional vector is added to the token's embedding vector. The resulting representation encodes both the semantic meaning of the token and its position in the sequence. However, the transformer model is expected to *implicitly* learn the sinusoidal pattern to understand token positions. No explicit structural signal is embedded to make this process easier or more interpretable for the model.

It is important to note that sinusoidal positional encoding is applied directly to the token embedding vectors, even though positional information is only required within the attention mechanism.

RoPE addresses these limitations with three key design goals:

1. Embedding the relative position of each token instead of using an absolute vector. This enhances the model's awareness of the token's position *relative to others* in the sequence.

2. Encoding the relative distance between tokens, ensuring that tokens separated by the same distance are rotated by the same angle.

3. There is no need to modify the embedding vectors after encoding the positional information. It is only used *on-the-fly* during the construction of the query and key vectors.

RoPE enables the model to better preserve and use distance and order information in the input, resulting in a deeper and more robust understanding of the sequence structure. Notably, RoPE achieves this without introducing any additional trainable parameters.

RoPE operates by rotating each pair of consecutive elements in the embedding vector by an angle proportional to the token's position. Specifically, for a token at position m, each pair is rotated by an angle of $m \times \theta$.

The rotation angle θ is computed using the formula:

$$\theta_i = 10000^{\frac{2i}{d}} \tag{1}$$

Where:

- d is the length of the embedding vector.

- i is an index ranging from 0 to $(d/2) - 1$, referring to each pair of dimensions.

For an embedding vector:

$$x = \left[x_0, x_1, x_2, \ldots, x_{d-1}\right] \tag{2}$$

RoPE forms consecutive pairs of dimensions:

$$\text{Pairs} \Rightarrow \left[(0,1),(2,3),(4,5),\ldots,(d-2,d-1)\right] \tag{3}$$

These are grouped into 2D vectors. The number of total groups is half the length of the embedding vector.

$$x \Rightarrow \left[(x_0,x_1),(x_2,x_3),(x_4,x_5),\ldots,(x_{d-2},x_{d-1})\right] \tag{4}$$

Each pair is rotated using a 2D rotation matrix based on sine and cosine functions:

$$\begin{bmatrix} x_m'^0 \\ x_m'^1 \end{bmatrix} = \begin{bmatrix} \cos(m\theta) & -\sin(m\theta) \\ \sin(m\theta) & \cos(m\theta) \end{bmatrix} \begin{bmatrix} x_m^0 \\ x_m^1 \end{bmatrix} \tag{5}$$

RoPE is not applied directly to the embeddings in memory. In other words, we are not interested in saving the new embedding vectors after the position encoding. Instead, the positional rotation is applied *on-the-fly* during the transformation into query and key vectors. This is why the query and key weights matrices are included in the equation:

$$\begin{bmatrix} \{q,k\}_m^0 \\ \{q,k\}_m^1 \end{bmatrix} = \begin{bmatrix} \cos(m\theta) & -\sin(m\theta) \\ \sin(m\theta) & \cos(m\theta) \end{bmatrix} \begin{bmatrix} W_{\{q,k\}}^{(11)} & W_{\{q,k\}}^{(12)} \\ W_{\{q,k\}}^{(21)} & W_{\{q,k\}}^{(22)} \end{bmatrix} \begin{bmatrix} x_m^0 \\ x_m^1 \end{bmatrix} \tag{6}$$

This can be generalized as:

$$f_{\{q,k\}}(x_m, m) = R_{\theta,m}^d\left(W_{\{q,k\}}(x_m)\right) \tag{7}$$

Where:

- [OBJ] is the rotary function applied to either the query or key vector at position m [OBJ].

- $W_{\{q,k\}}$ is the corresponding weight matrix.

- $R^d_{\theta,m}$ is the rotation operator, which generates $d/2$ individual rotation matrices each of size 2×2.

- x_m is the embedding vector of the token at position m.

Figure 3-1 shows how the embedding vectors of the tokens *sky* and *bright* are rotated in the embedding space after applying RoPE. Each new vector results from a rotation applied to the original embedding vector, preserving the relative positional relationship between the two tokens.

Figure 3-1. *Illustration of rotating the embedding vectors of the tokens sky and bright using RoPE*

As long as the distance between the corresponding tokens remains the same within the sequence, RoPE can preserve the relative angles between embedding vectors. As shown in Figure 3-2, even if token positions shift after introducing the new word *yesterday*, the relative distance and angular relationship between the vectors stay consistent. This allows the model to more effectively capture and leverage relative positional information.

Figure 3-2. *The two embedding vectors maintain the same relative angle if the corresponding tokens are separated by the same distance in the sequence*

Due to the efficiency of RoPE, it is used in Meta's LLaMA 1, 2, 3, and 4, DeepSeek V3 and R1, Google PaLM, etc.

3.2 Other Attention Techniques

In the traditional multi-head attention (MHA), each head uses different sets of weights for these matrices:

1. key

2. value

3. query

If the LLM uses $h = 16$ heads and an embedding size of $d_{model} = 2,048$, then the key weights dimension is $d_k = \dfrac{2,048}{16} = 128$. Consequently, each key projection matrix for a single head has a size of $2,048 \times 128$. The same applies to the query and value projection matrices for a total of $16 \times 3 = 48$ weight matrices.

These projection matrices contribute a significant number of parameters that must be computed during inference for every token. Even though their matrix multiplications can be parallelized, the computations are still repeated for each generated token. Additionally, storing a separate set of projection matrices for each head increases memory usage. As the context window size increases in modern LLMs, MHA becomes one of the main computational and memory bottlenecks.

We will examine three variations of MHA designed to overcome its limitations.

1. Multi-query attention

2. Grouped-query attention

3. Multi-head latent attention

3.2.1 Multi-Query Attention (MQA)

While the original Transformer architecture set the stage for the generative AI revolution, its standard multi-head attention mechanism introduces a significant computational bottleneck during inference.

In large-scale deployments, the need to store and retrieve a unique Key and Value (KV) cache for every single attention head across every layer leads to massive memory consumption and high latency. As models scale to billions of parameters, this KV cache often becomes the primary limiting factor for both the speed of text generation and the number of users a single server can support.

MQA was introduced in the paper titled "Fast Transformer Decoding: One Write-Head Is All You Need" by Noam Shazeer to optimize the performance of LLMs. The key idea behind MQA is to share the same key and value weight matrices across all the attention heads as in Figure 3-3. This approach offers two main advantages:

1. Reduced memory usage by storing a single set of key and value weights instead of separate ones for each head.

2. Lower computational cost by performing the matrix operations once and reusing the results across multiple attention heads.

Instead of using $16 \times 3 = 48$ weights matrices, MQA only used $1(\text{shared key}) + 1(\text{shared value}) + 16(\text{queries}) = 18$ matrices. This is a huge reduction.

Each of the N query matrices is multiplied by the shared key matrix to produce N intermediate attention scores. As in Figure 3-3, there are eight heads that generate eight intermediate outputs when multiplied by the shared key matrix. As usual, scaling and softmax function are applied to each output. Each of these is then used to compute an output by multiplying it with the shared value matrix to produce 8 outputs. The resulting outputs are finally concatenated to form the complete attention output.

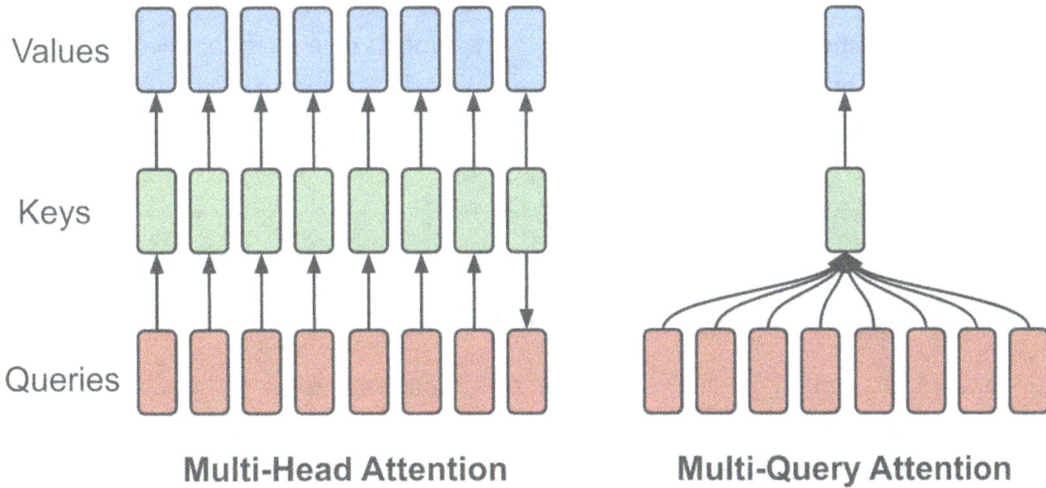

Figure 3-3. *A comparison between the multi-head attention (MHA) and the multi-query attention (MQA). MHA uses a different key and value weight matrix for each transformer attention head. While MQA uses the same weight matrices across all heads*

The multi-query attention reduces the number of parameters dramatically compared to the multi-head attention. But this comes at the cost of reducing the capacity of the model when measuring the attention score between the tokens. Reducing the trainable parameters in a model reduces its capacity. Instead of being able to digest a certain amount of information at a time, this capacity will decrease.

Even that one of the objectives of the MQA is reducing memory consumption by storing shared weights matrices for each head; this is not completely possible due to the parallel calculations in the transformer. To apply such operations in parallel, there is a need to replicate the same key and value matrices across all the nodes. But still MQA is much memory efficient than the traditional MHA.

If the goal of MQA is to reduce the number of parameters, a natural question arises: why not share the query weight matrix across all heads as well? Although being highlighted earlier in the explanation of MHA, this is worth examining. Drawing an analogy from information retrieval systems, the query represents the question or search string used to retrieve relevant information (e.g., searching for a video on YouTube). If the same query is used repeatedly, it will return the same type of information every time.

In the context of LLMs, sharing the query matrix across all heads would result in all heads returning the same outputs. This defeats the core purpose of using multiple heads in the Transformer's attention mechanism. Each head is intended to focus on different aspects of the relationship between the tokens by using its own query projection. When each head has a unique query matrix, the model is encouraged to explore diverse relationships among the tokens. This diversity allows the model to learn richer representations, enabling it to determine which tokens are more relevant to each other and, ultimately, which tokens should attend to which.

3.2.2 Grouped-Query Attention (GQA)

MHA is computationally expensive and memory-intensive because it uses a separate key and value weight matrix for each attention head. However, this design preserves the full expressive capacity of the model, leading to higher accuracy in generating tokens.

MQA improves efficiency by sharing the key and value weight matrices across all heads. This reduces both memory usage and computation time, especially for long sequences. However, this simplification comes at the cost of accuracy, as the model loses much of the trainable parameter space within the attention block. As a result, MQA may struggle to accurately determine which tokens should attend to each other.

To balance efficiency and performance, Grouped-Query Attention (GQA) offers a middle-ground solution. It is introduced in 2023 in the paper titled "GQA: Training Generalized Multi-Query Transformer Models from Multi-Head Checkpoints."

Instead of sharing key and value weights across all heads, GQA shares them within smaller groups of heads. For example, in a configuration with 8 heads and 2 groups, each group of 4 heads shares one key and one value weight matrix. This is illustrated in Figure 3-4. This maintains more diversity across the attention heads while still reducing the computational and memory requirements compared to standard MHA.

Note that both MQA and GQA do not group the queries across the attention heads.

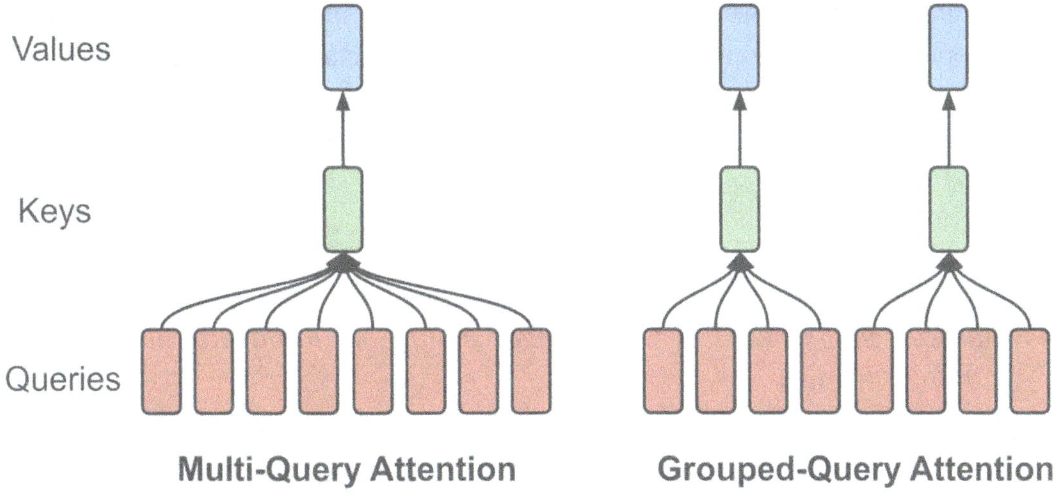

***Figure 3-4.** Comparison between multi-query attention (MQA) and grouped-query attention (GQA). GQA groups the weights matrices of multiple keys and values across multiple heads*

The N query matrices are divided into G groups, where each group shares a single key and value matrix. Within each group, the query matrices are individually multiplied by the shared key matrix to compute intermediate outputs. In Figure 3-4, the group has 4 heads and thus the multiplication with the key matrix produces 4 intermediate matrices. Each matrix is applied to scaling and softmax function. These intermediate matrices are then multiplied by the shared value matrix to produce 4 outputs. Such outputs from all groups are then concatenated to produce the final attention output.

GQA offers more efficient memory usage compared to MQA by enabling a degree of parallelism. Since each group of attention heads in GQA operates independently, different groups can be processed simultaneously across multiple compute nodes.

Compared to MQA, which typically requires training a model from scratch to achieve strong performance, GQA can be applied to pretrained models that already used MHA. This is done through a fine-tuning process known as *uptraining*.

Unlike conventional fine-tuning which adapts a model to a new task or dataset, uptraining involves modifying the model's architecture while retaining its original parameters. In this case, the architecture is updated to incorporate GQA, making the model a more efficient version of itself. Uptraining is significantly less complex and resource-intensive than training a model from scratch, especially when combined with parameter-efficient techniques such as Low-Rank Adaptation (LoRA).

Due to its balance between parameter efficiency and computational speed, GQA has been adopted in several popular large language models, including Meta's LLaMA 2 and LLaMA 3, Mistral 7B by Mistral AI, and IBM's Granite 3. However, some models continue to use standard MHA such as OpenAI's GPT-3 and GPT-4 to preserve the full capacity of the model.

It is important to note that applying MQA or GQA is not as simple as modifying a trained model's architecture to share key and value weight matrices across heads. These changes should be introduced during training, not after.

If a model is trained with standard MHA and then switched to MQA or GQA during inference, its performance will degrade. This is because the model was originally optimized to use the full capacity of the attention mechanism and reducing that capacity post-training disrupts the learned representations. The same principle applies to other optimization techniques such as quantization. Quantizing a model after training can also result in degraded performance. For best results, these efficiency improvements must be either incorporated into the training pipeline or followed by a fine-tuning phase that adapts the model to the modified architecture or numerical precision.

3.2.3 Multi-Head Latent Attention (MLA)

Multi-Head Latent Attention (MLA) is a technique introduced in the paper titled "DeepSeek-V2: A Strong, Economical, and Efficient Mixture-of-Experts Language Model" by the DeepSeek team in May 2024. It is used in both DeepSeek-V2 and DeepSeek-V3 models.

MLA is designed to reduce computational and memory costs during inference while maintaining model expressiveness. Unlike Grouped Query Attention (GQA), which shares key and value matrices across groups of attention heads, MLA maintains separate key and value matrices for each head but introduces architectural changes to achieve significant efficiency gains. One of the primary motivations behind MLA is to reduce the size of the key-value (KV) cache, which is a major bottleneck in autoregressive models.

Instead of generating the full key and value matrices in a single step using high-dimensional projection, MLA breaks the process into two stages:

1. **Down-Projection to Latent Space**: The token embedding is projected into a lower-dimensional latent space using a down-projection matrix. Only these compressed representations are cached, resulting in a substantial reduction in KV cache size.

2. **Up-Projection When Needed**: When computing the attention
    scores, the cached latent vectors are up-projected back to full-
    sized key and value vectors only as needed.

This allows the model to apply the attention mechanism by using different key and value vectors for each attention head while avoiding storing high-dimensional key and value vectors for every past token. Instead of caching the full embeddings, MLA caches their compressed latent representations.

Let h be the token's hidden state vector (e.g., the output of the previous layer). The query vector is compressed using a down-projection matrix W^{DQ}, where D stands for *down-projection* and c refers to a compressed version:

$$c^Q = W^{DQ} h \tag{8}$$

Similarly, the key and value are compressed using a shared down-projection matrix W^{DKV}:

$$c^{KV} = W^{DKV} h \tag{9}$$

Only the compressed c^{KV} is cached, since key and value vectors are reused across all future tokens. In contrast, the query vector is computed fresh for each token, so it is not cached.

To recover the full-resolution vectors needed for attention, up-projection matrices are used:

- For the query:

$$q^C = W^{UQ} c^Q \tag{10}$$

- For the key and value:

$$k^C = W^{UK} c^{KV}$$
$$v^C = W^{UV} c^{KV} \tag{11}$$

Here, U stands for *up-projection*. These restored vectors are then used in the standard scaled dot-product attention mechanism. Figure 3-5 explains how the MLA works.

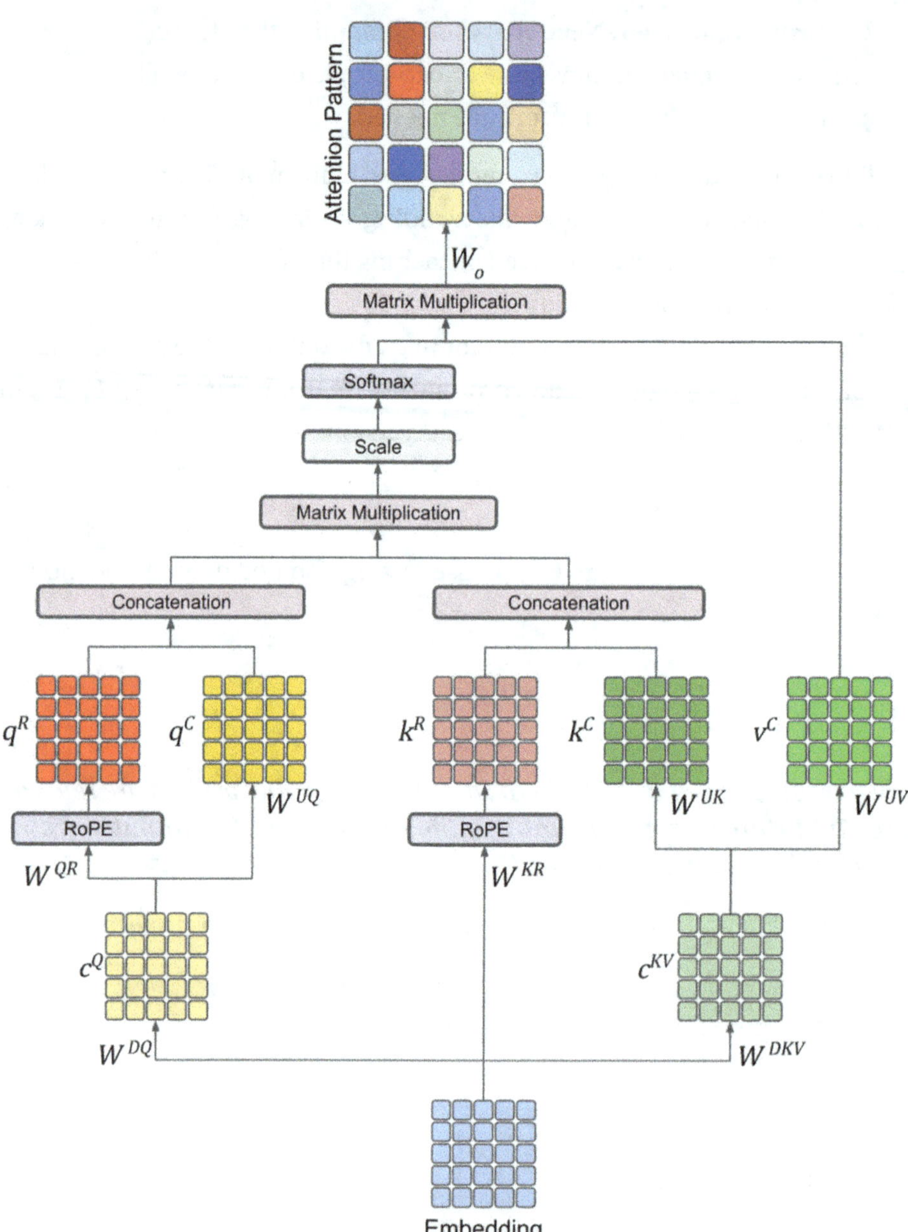

Figure 3-5. *Multi-head latent attention (MLA) projects the key, query, and value representations into a lower-dimensional latent space to reduce the size of the key–value (KV) cache. When required, up-projection vectors are applied to restore these representations to their original dimensionality. Once the full-size matrices are reconstructed, the attention mechanism proceeds as in the standard transformer architecture. Rotary Position Embedding (RoPE) is applied to preserve positional information throughout the process*

The MLA mechanism creates two latent spaces:

1. One for queries

2. One shared for keys and values

As illustrated in Figure 3-5, each down-projected latent vector in the query latent space, denoted c^Q, is passed through two operations. The first, previously discussed, involves up-projecting the query latent vector. The second operation is the application of RoPE (Rotary Positional Embedding) to encode positional information. This is computed using the following equation:

$$q^R = RoPE\left(W^{QR}c^Q\right) \tag{12}$$

Here, W^{QR} is the transformation matrix that enables RoPE to rotate the query latent vector c^Q based on positional information. Similarly, RoPE is applied to the token embedding vector (not the latent vector) to represent the positional information of the key after up-projection.

Applying RoPE is essential for preserving token position information after compression into a latent space. Once the embedding vector is compressed, any position encoding it previously contained is lost. To restore this information, RoPE is applied to reintroduce positional structure during the up-projection phase. RoPE works by rotating a vector by an angle determined by the token's position in the sequence.

After obtaining the RoPE-enhanced vector and the up-projected vector, these are concatenated for both the query and the key. This concatenation produces a composite vector that combines both content and positional information. Standard attention operations then follow: matrix multiplication, scaling, softmax, and multiplication with the value vector. The value vector is derived by applying the up-projection to C^{KV} using the weight matrix W^{UV}.

This is the final equation for the MLA:

$$o_{t,i} = \sum_{j=1}^{t} \text{Softmax}_j\left(\frac{q_{t,i}^{\top}k_{j,i}}{\sqrt{d_h + d_h^R}}\right)v_{j,i}^C \tag{13}$$

Where the indices refer to:

- t: The current time step or position in the sequence for which the output is being computed.

- j: The position in the sequence being attended to. It ranges from 1 to t, so this is causal (autoregressive) attention where only past and current positions are considered.

- i: The attention head index, since multi-head attention runs multiple attention mechanisms in parallel.

- d_h: The dimensionality of the content (original) part of the query and key vectors (i.e., after up-projection).

- d_h^R: The dimensionality of the RoPE part that is concatentated to inject positional information.

Because the length of the output key and query vectors becomes the length of the embedding vector d_h and the RoPE vector d_h^R, then scaling is applied by the sum of these two lengths. As usual, the output of the attention block is computed by multiplying o_t by the output projection matrix W^o.

As shown in Figure 3-5, RoPE is applied to the token embedding vector to encode position for the key, not to the key's latent vector. In contrast, for the query, RoPE is applied directly to the latent query vector. This design choice is motivated by two reasons:

1. RoPE is applied over the smaller, compressed query vector c^Q, rather than the full embedding vector h, making it more efficient.

2. For the key, there is no dedicated latent vector. Instead, the key and value share the same latent vector c^{KV} to reduce memory usage in the key-value (KV) cache. Since c^{KV} encodes both key and value information, it is unsuitable for RoPE application via a dedicated key transformation matrix W^{KR}.

3.2.4 Sliding Window Attention (SWA)

The original attention mechanism computes attention scores between each token in the context and every other token, using full-sized key, query, and value matrices for the attention heads. Efficiency improvements were introduced by approaches such as multi-query attention and grouped-query attention, which share parameter matrices across heads, and multi-head latent attention, which reduces the dimensionality of the matrices.

With the advent of recent models, context lengths have expanded dramatically, reaching up to one million tokens in Google Gemini 2.5. At such scales, computing attention between every token pair becomes very time-consuming and memory intensive.

Sliding-window attention (SWA) addresses this limitation by restricting attention computation to tokens near the target token. In most natural language tasks, a token is more closely related to its immediate neighbors than to distant tokens. Instead of using the full global context, SWA creates a local context window of size w tokens, where the target token attends to approximately $\frac{1}{2}w$ tokens to its left and $\frac{1}{2}w$ tokens to its right, as illustrated in Figure 3-6 where $w = 4$. In SWA, each token influences and is influenced only by the tokens within its predefined attention window by contributing to the embedding vectors of those nearby tokens while ignoring distant ones.

Figure 3-6. *Sliding window attention of a fixed window of $w = 4$ tokens. For each token, the model attends to approximately half of the tokens (2) within the window on its left and half on its right*

In the regular self-attention mechanism for causal tasks, an attention matrix of size $N \times N$ is generated, as shown on the left side of Figure 3-7. Because the task is causal, future tokens must be masked to prevent the model from accessing them during prediction. These masked positions are represented by zeros in the matrix. As a result, each token attends to all previously seen tokens (including itself). For example, token D attends to the three preceding tokens (A, B, and C), while token E attends to all four previously processed tokens (A through D) to its left.

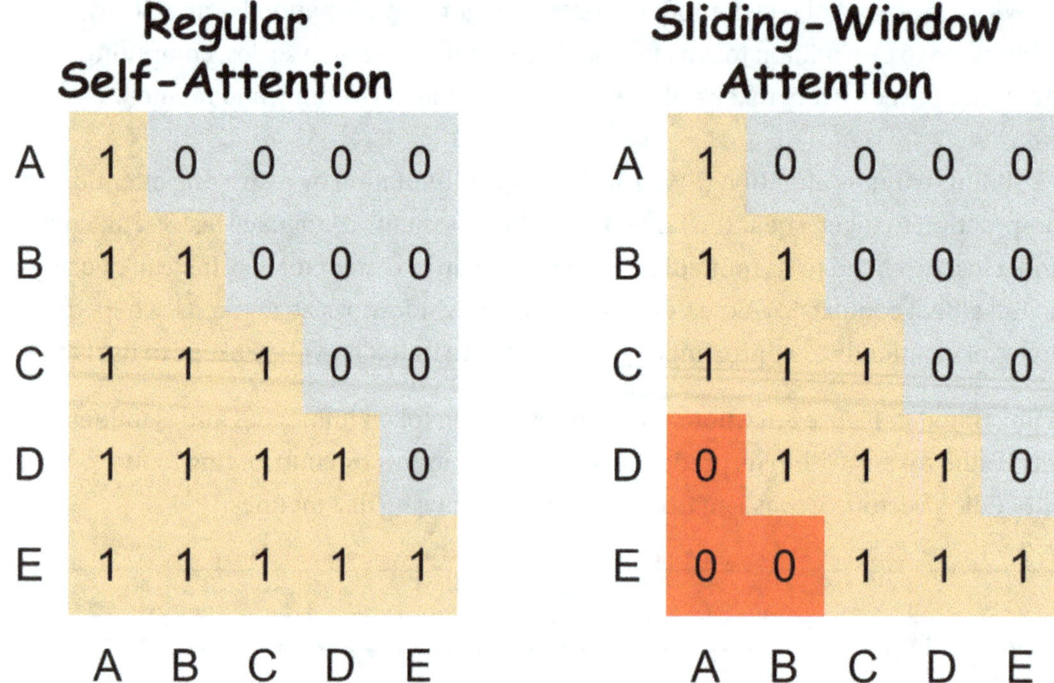

Figure 3-7. *Traditional causal self-attention versus sliding-window attention. In traditional self-attention, each token attends to all previously seen tokens. In contrast, sliding-window attention restricts attention to a fixed local window, attending only to tokens within that window to reduce computation*

In SWA, the scope of attention is reduced by restricting each token to attend to a fixed number w of surrounding tokens. For example, if $w = 4$, the window spans two tokens to the left and two tokens to the right of the target token. In a causal setting, the tokens to the right represent future (unseen) positions and are therefore masked, leaving only a maximum of $\frac{w}{2} = 2$ tokens to the left available for attention.

For instance, when processing token D, there are already three tokens to its left (A, B, and C). Since SWA can attend to at most two tokens to the left, only the most recent ones (B and C) are considered. These are the tokens closest in position to the current token D. Similarly, by the time token E is processed, four tokens have already been visited (A through D). As a result, SWA attends only to the latest two (C and D).

Although tokens most often attend to their immediate neighbors, there are cases where two related tokens fall outside the local attention window. This limitation is mitigated in deeper layers of the transformer, where successive applications of sliding-window attention allow information to propagate gradually across the sequence. In this way, higher-level layers build upon the localized attention patterns of earlier layers, enabling the model to capture long-range dependencies and develop a form of global awareness despite the local restriction in each layer. This process is similar to how a convolutional neural network (CNN) analyzes an image using small sliding windows, as illustrated in Figure 3-8.

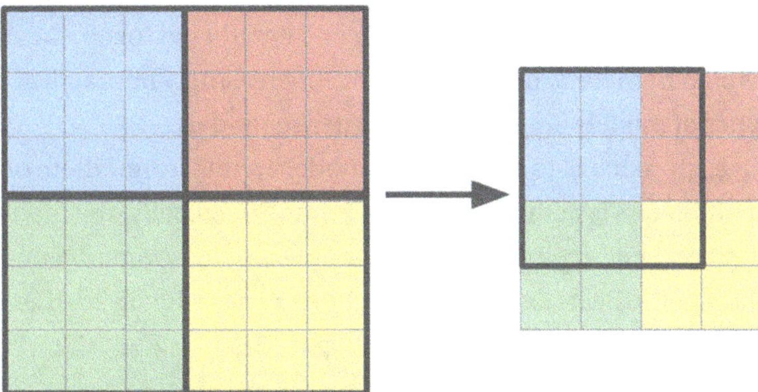

Figure 3-8. *Convolutional neural networks process inputs using sliding windows similar to the sliding-window attention mechanism in transformer models. In CNNs, each sliding window covers a small local region of the input to extract features, which are then combined across layers to capture larger patterns and develop a global understanding of the data*

Even though the window sizes (such as 3×3 or 5×5) are tiny compared to the full image size (e.g., 1024×1024), the network can still capture the global characteristics of the image and identify regions containing objects such as human faces or vehicles. This is because the early layers use these windows to extract local features. As the network goes deeper, the receptive fields expand by combining outputs from previous layers' windows.

In the right part of Figure 3-8, the new 3×3 window processes outputs from four different windows shown in the left part, enabling the network to move beyond purely local features and capture broader patterns. These progressively larger receptive fields capture relationships across different local regions, enabling the network to develop a global understanding of the entire image.

3.3 Mixture of Experts (MoE)

In traditional machine learning models, the entire capacity of the model is typically utilized during both training and inference. Data samples propagate through the full architecture in the same way for both phases. While this approach may be acceptable during training, it presents a significant bottleneck during inference, especially for LLMs, which often contain billions of parameters. For each predicted token, the model must compute across all parameters, making inference increasingly impractical as model size grows. These types of models are referred to as dense models.

An alternative approach is to use a sparse model architecture, where only a subset of the model's parameters is activated for each input. Specifically, the model is designed with multiple paths called "experts," and each input sample traverses only a small number of these experts. Because there are multiple experts involved, this technique is referred to as a Mixture of Experts (MoE), as in Figure 3-9.

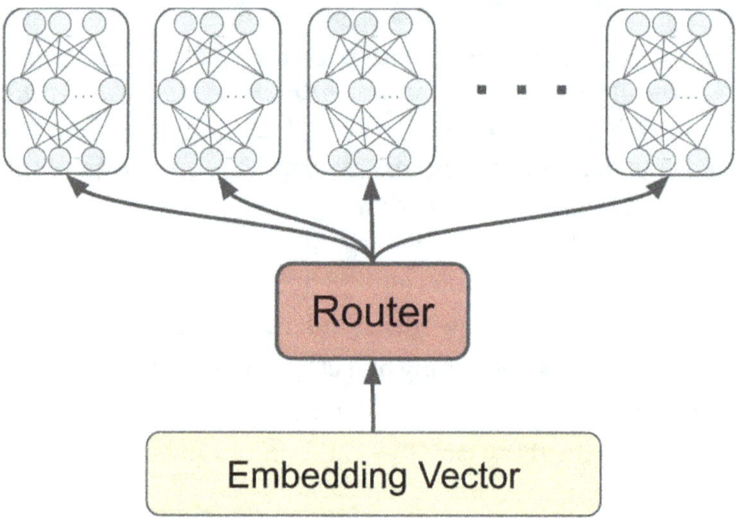

Figure 3-9. *Mixture-of-experts (MoE) replaces a single large feed-forward neural network with multiple smaller expert networks. For each input token, only a subset of these experts is activated and used for computation. This sparse activation increases model capacity while maintaining computational efficiency*

Each expert model is a feed-forward neural network, trained on a particular subset of the data. These experts can be thought of as specialists analogous to people with expertise in distinct domains such as medicine, finance, law, or engineering. However, in generative AI systems, experts are often not organized by topical domains but rather by linguistic or structural characteristics of the input. For example, some experts may specialize in handling long-range dependencies, while others focus on syntactic structure, rare vocabulary, multilingual text, dialogue formatting, or programming languages.

Instead of relying on a single model with, for example, 10 billion parameters, the architecture may consist of 10 different experts, each with a smaller number of parameters (e.g., 1 billion). In this case, each input only uses a few of these experts, reducing both training and inference costs. Assuming that each token is passed to three experts, the inference time can be approximately 3/10 of what would be required if the full model were used, without necessarily sacrificing performance. This efficiency gain applies to training time as well, since fewer parameters are updated per input.

One of the key advantages of the MoE approach is that it enables a model to have high overall capacity while maintaining computational efficiency during training and inference. Consider a model with 10 expert networks, each containing 2 billion parameters. Although the total parameter count is doubled, the MoE is designed to activate only a subset of experts, for example, 2 experts per input.

In this case, for any given input token or example, only 4 billion parameters (2 experts × 2 billion each) are actually used. This selective activation means that while the model has access to a large parameter space (increased capacity), it only processes a fraction of the parameters per input. As a result, the model remains efficient in terms of memory and computation, achieving performance benefits similar to those of a large dense model but at a much lower cost.

A critical component of MoE architectures is the router, a mechanism that analyzes the input and determines which experts to activate. The router itself is usually represented as a single dense layer connected to a softmax function to generate probabilities of selecting each expert to analyze the token. The higher the probability, the more aligned is the token with the expert network.

Once the router selects the experts for a given input, the softmax probabilities are used to aggregate the outputs of the active experts into a final result, as in Figure 3-10, where only 2 experts are selected. This weighted combination allows the model to produce a coherent prediction that leverages the strengths of the most relevant expert models.

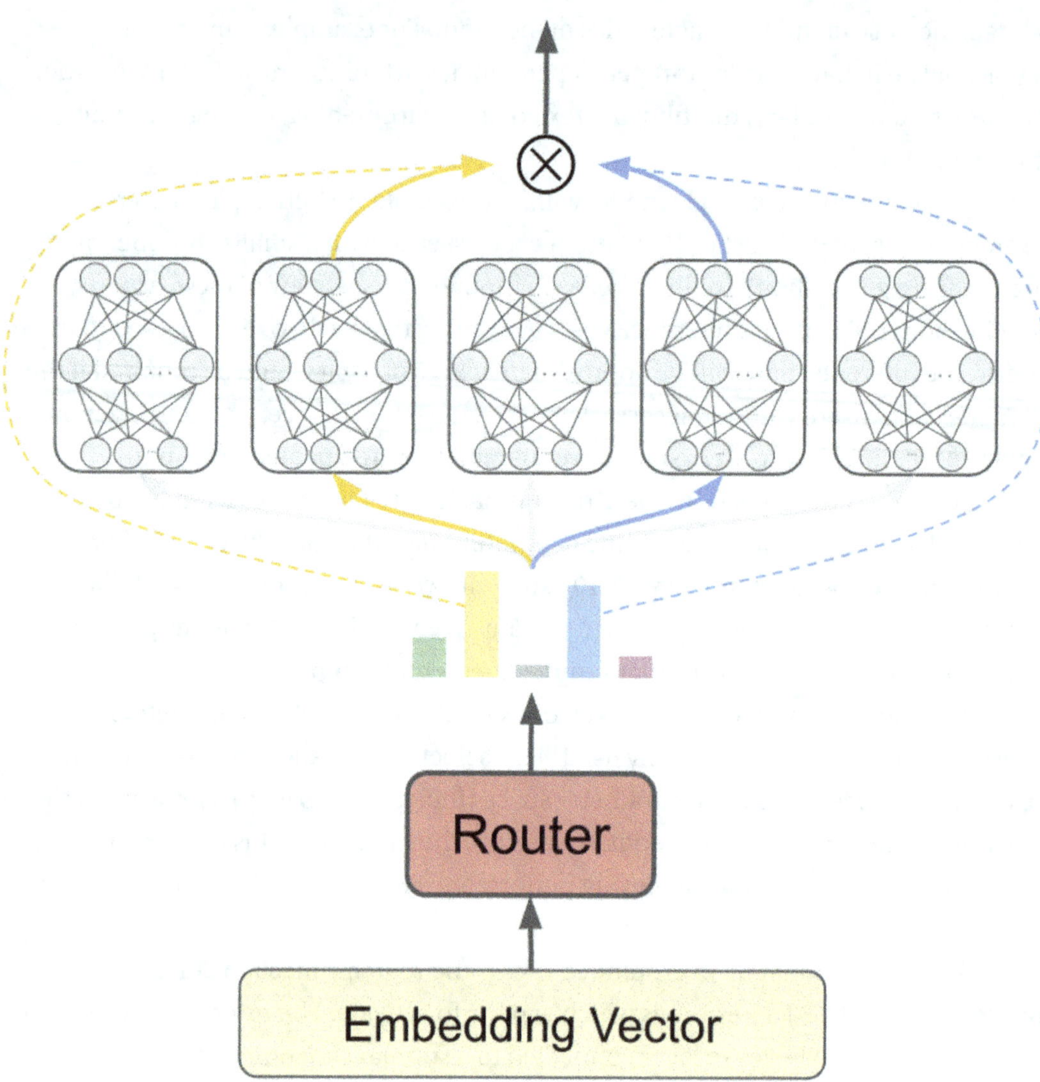

Figure 3-10. *Illustration of the Mixture-of-Experts (MoE) model router. A router assigns a weight to each expert based on a softmax function applied to routing scores. The top N experts with the highest weights are selected to process the input embedding vector. Their outputs are then combined, weighted by the corresponding softmax scores, to produce the final representation*

The router must be carefully trained to perform two key functions:

1. Routing inputs to the most appropriate expert(s) based on the content or structure of the input.

2. Ensuring load balancing across all available experts to optimize resource utilization and maintain model efficiency.

The router is responsible for selecting the most appropriate expert(s) for each input token or segment. If the router fails to direct inputs to suitable experts, those inputs may be processed by experts that are ill-suited to handle them, which can negatively impact the model's performance.

In addition to selecting the correct experts, the router must also ensure load balancing across them. This is essential to avoid overloading a small subset of experts while others remain underutilized. MoE models rely on parallelism by distributing inputs across multiple experts. If the router consistently favors only a few experts, the model's parallelism, scalability, and overall efficiency are diminished. Balanced routing helps preserve these benefits and contributes to better generalization during training. One way to regularize routing and increase the probability of selecting less likely experts is to introduce Gaussian noise to the routing logits.

In traditional MoE architectures, each token is typically routed to a fixed number of experts, regardless of the token's semantic importance or complexity. This uniform routing strategy can be suboptimal, as not all tokens contribute equally to the model's output because some may be more important than others.

To address this limitation, researchers at Google proposed a novel routing method called "Expert Choice," as in Figure 3-11, introduced in the paper titled *"Mixture-of-Experts with Expert Choice Routing."* Unlike standard routing, which assigns each token to a fixed number of experts, Expert Choice performs token-to-expert routing from the expert's perspective. That is, each expert selects which tokens to process, allowing the number of experts handling each token to vary based on token importance.

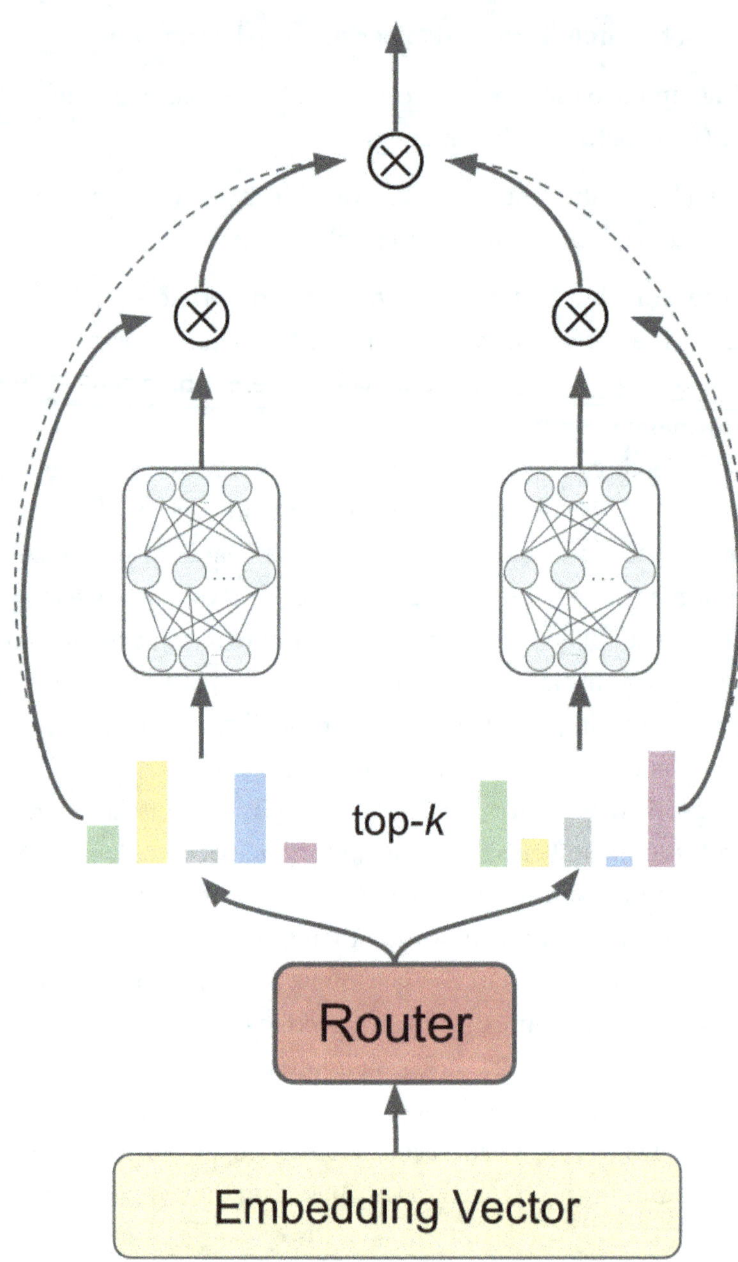

Figure 3-11. *Routing in a mixture-of-experts (MoE) model with expert choice, where each expert is assigned the top K tokens based on their routing importance scores. This approach ensures that each expert processes only the most relevant tokens*

Figure 3-11 assumes that there are five input tokens, as indicated by the number of weights for the top tokens. For each expert, the router generates weights for all tokens, but only the top K tokens are fed to the expert. The outputs of all experts are aggregated using the generated weights to create the new embedding vector.

In standard routing, the top-k experts receive a token.

In Expert Choice, the expert receives the top-k tokens.

To prevent experts from becoming overloaded with tokens, the approach introduces a parameter known as expert capacity. This defines the maximum number of tokens an expert is allowed to process in a given batch. Once an expert reaches its capacity, it rejects additional tokens. This constraint maintains computational efficiency and ensures a more balanced load across experts.

Due to the efficiency of the MoE in LLMs, it is used by some models such as Mixtral 8x7B by Mistral AI, DeepSeek V3 by DeepSeek AI, Qwen3 235B, LLaMA 4 by Meta, Qwen2.5-Max by Alibaba, etc.

LLaMA 4 Maverick uses a shared expert in its MoE architecture, which is applied to all tokens, in addition to a token-specific selected expert as in Figure 3-12. This model has a total of 128 routed experts in addition to the shared expert. The total number of parameters in this model is 400 billion, but only 17 billion are activated for each token. It is a massive model that supports 1 million tokens. LLaMA 4 Scout increased the context length to 10 million tokens while only using 16 routed experts. It also has 17 billion parameters activated at a time with a total of 109 billion parameters.

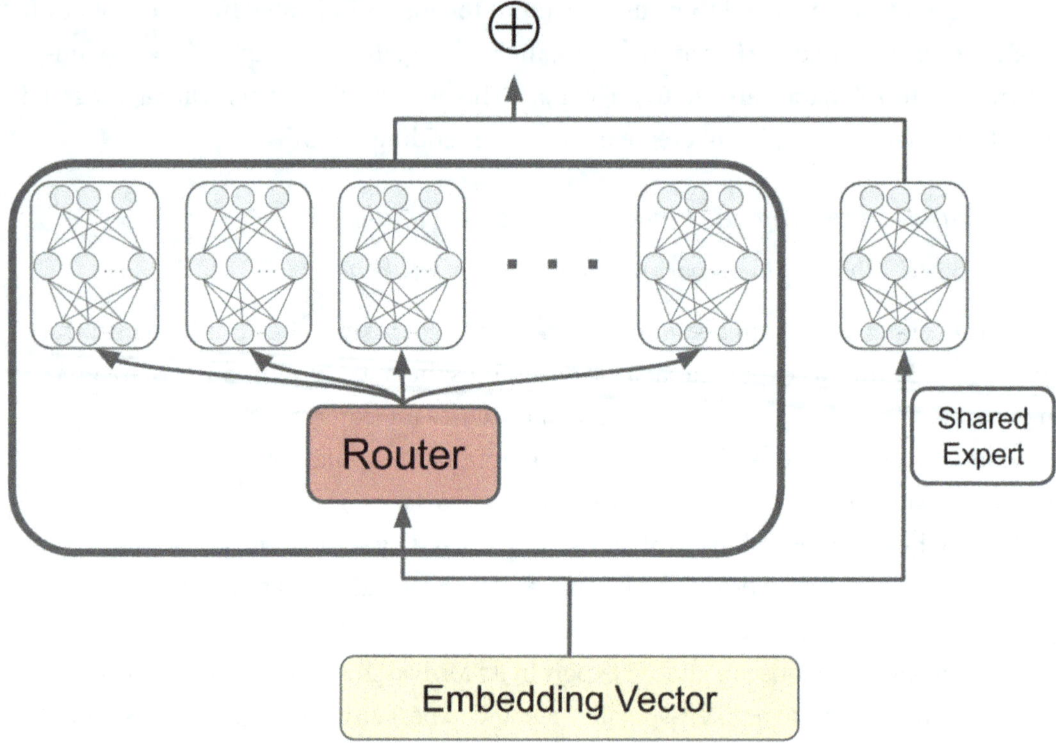

Figure 3-12. *LLaMA 4 by Meta incorporates a shared expert that is applied to all tokens, alongside a dynamically selected expert determined on-the-fly by a routing mechanism*

The motivation for including a shared expert is that it can learn general linguistic patterns that are broadly applicable across all tokens. By combining the output of the shared expert with that of the dynamically selected experts, the model captures both general and specialized token-level information, leading to improved representation quality and better performance in the attention mechanism.

3.4 Small Language Model (SLM)

An LLM is considered "large" due to its extensive architecture and the massive amount of data used during training. LLMs typically contain hundreds of billions of parameters. Developing such a model from initial research to a fully trained system capable of understanding and generating natural language can take several months and requires substantial computational resources.

Regarding an autoregressive LLM, it generates text one token at a time in a sequential manner. To produce each token, the model processes an input embedding that includes the original prompt along with any previously generated tokens. This entire sequence must pass through the full model architecture, making the generation process computationally intensive.

LLMs are usually trained on broad, diverse datasets, enabling them to develop general knowledge across a wide range of domains, such as healthcare, sports, e-commerce, and technology. However, deploying general-purpose LLMs in real-world applications is not always practical due to several limitations:

1. **Task Specialization**: Many applications require domain-specific expertise, making a general-purpose model inefficient or unnecessary.

2. **High Inference Cost**: Running inference with LLMs is both resource-intensive and time-consuming. This introduces latency that is unsuitable for real-time applications like customer support. Furthermore, performance can degrade if the model needs to access external tools or APIs during inference.

3. **Maintenance Challenges**: Updating or fine-tuning LLMs with new data is costly and complex. In contrast, smaller models are easier to retrain and maintain.

For example, a cybersecurity enterprise may use a language model to analyze device logs for signs of vulnerabilities or risks. Given the narrow focus of this task, using a general-purpose LLM would be inefficient. A smaller, task-specific model optimized for cybersecurity analysis would be more effective and cost-efficient.

Small language models (SLMs) are another class of models that are significantly smaller than traditional LLMs. Typically ranging from hundreds of millions to a few billion parameters while being trained on domain-specific data. Their smaller size makes them faster to train, easier to deploy, and more cost-effective to operate.

Examples of SLMs include:

- DistilBERT—Hugging Face

- Phi-1.5 and Phi-2—Microsoft

- MiniLM—Microsoft

- LLaMA 3-8B—Meta

- Gemma-2B—Google

- Mistral-7B—Mistral AI

In general, SLMs can be categorized based on the strategy used to reduce model size while maintaining acceptable performance. The main strategies include:

1. **Distilled**: This approach uses a larger LLM (*teacher*) to train a smaller model (*student*) by mimicking the teacher's outputs. The student model learns to approximate the teacher's behavior, often resulting in a model that is significantly smaller but still has a competitive performance. Examples include DistilBERT and MiniLM.

2. **Lightweight General-Purpose**: Instead of compressing an existing model, some SLMs are designed from the ground up to be efficient. These models have some changes in their architecture, such as fewer layers or sparsity techniques to build general-purpose models that are inherently smaller and faster. Examples include Phi-2, Mistral-7B, and Gemma-2B.

3. **Task-Specific**: In this strategy, a smaller model is trained exclusively for a specific domain or task (e.g., sentiment analysis, medical text classification, or cybersecurity log parsing). By narrowing the scope, the model can achieve high accuracy without the need for billions of parameters. These models are often fine-tuned versions of other models, tailored for domain-specific applications.

3.4.1 Knowledge Distillation

Knowledge distillation (KD) is a technique that enables a smaller, more efficient student model to learn from a larger, often more accurate teacher model. The knowledge transferred may represent a new skill or understanding of a specific domain. In neural networks, this can include learning how to extract features similar to the teacher.

As illustrated in Figure 3-13, KD methods can be categorized based on the type of knowledge being transferred to the student model:

1. **Response-Based**: In this approach, the student learns to replicate the final output of the teacher model, typically the softmax probabilities over the output classes. The student is trained to match this output distribution, enabling it to generalize in a way similar to the teacher.

2. **Feature-Based**: The focus is on transferring intermediate representations from the teacher's internal layers, not the final output. The student model learns to extract features in a way that mimics specific hidden layers of the teacher. This is especially useful when the goal is to preserve the way the teacher transforms the input, not only matching the output predictions.

3. **Relation-Based**: Instead of preserving the behavior of a single layer, the student attempts to preserve the structural dependencies, such as correlations or distance metrics, between different layers or data points in the teacher model. For instance, the student may be trained to maintain the same relational structure between intermediate layer outputs and the final output. This is a more advanced strategy.

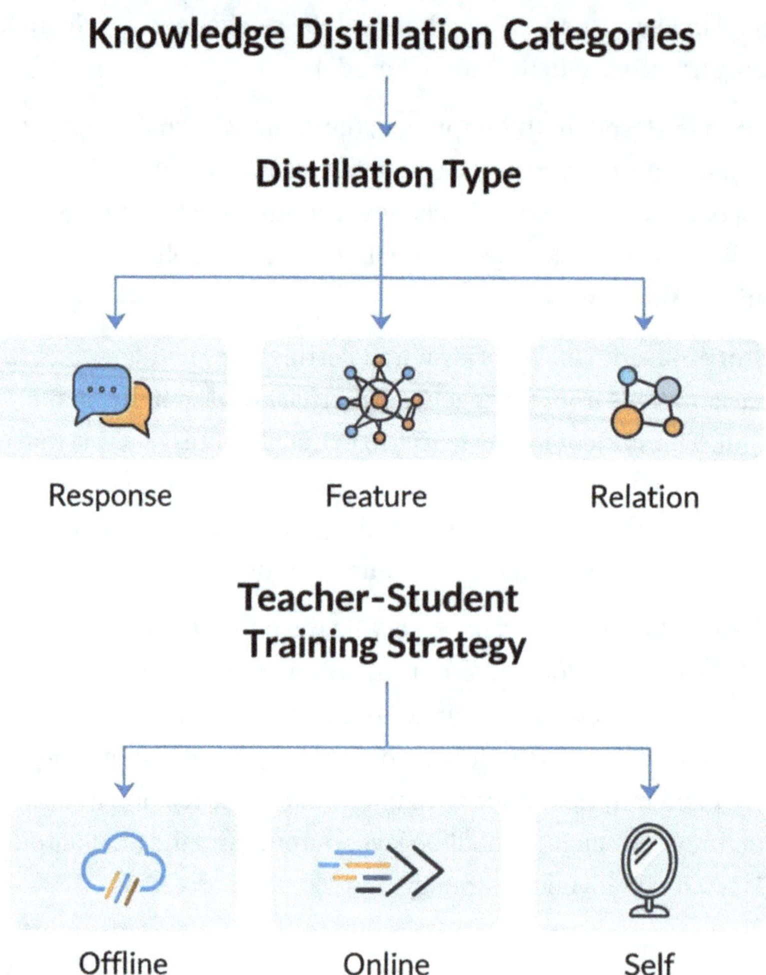

Figure 3-13. *Categorization of the knowledge distillation. Based on the distillation type, it could be either response, feature, or relation. Based on the training strategy, it is either offline, online, or self-distillation*

Figure 3-14 illustrates the general process of knowledge distillation. Regardless of the specific type of knowledge being transferred (response, feature, or relational), the student model is trained to absorb some form of learned knowledge from the teacher. In this context, knowledge refers to the trainable parameters (weights) that embody the teacher model's learned behaviors.

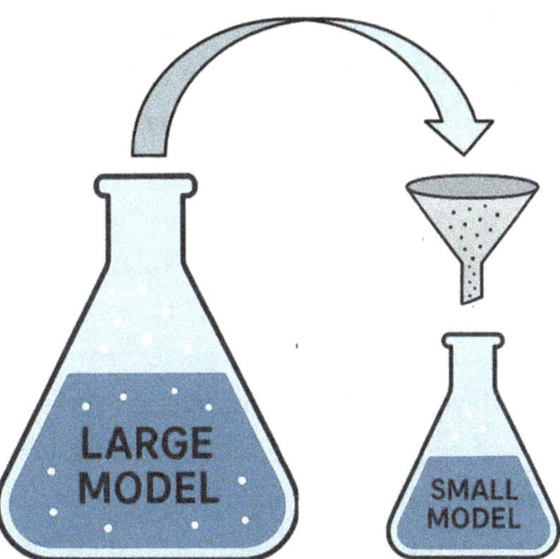

Figure 3-14. *Knowledge distillation is the process of transferring knowledge from a large, high-capacity model (the teacher) to a smaller, more efficient model (the student), enabling the student model to achieve comparable performance with reduced computational requirements*

While the distilled knowledge is often not as comprehensive as that of the original model, the student model can still achieve a high degree of performance, efficiently replicating many of the teacher's capabilities with significantly fewer parameters.

During the training of a student model, knowledge distillation can be categorized into three categories:

1. **Offline Distillation**: In this approach, a pre-trained teacher model is used to generate outputs (responses or predictions) that are then used to train the student model. The term *offline* refers to the fact that the teacher remains fixed (i.e., frozen) during the student's training process. No further updates are made to the teacher during distillation.

2. **Online Distillation**: In contrast, *online* distillation involves training both the teacher and the student models simultaneously. As the teacher improves during training, it provides increasingly refined outputs, allowing the student to learn progressively. This dynamic learning process enables the student model to benefit from the teacher's evolving knowledge, often resulting in better performance compared to the static offline setting.

3. **Self-Distillation**: In self-distillation, a single model serves as both the teacher and the student. Knowledge is distilled internally, specifically from the deeper layers to the shallower layers. This method enhances the representational power of the early layers and improves overall model performance. The paper that introduced self-distillation, titled "Self-Distillation: Towards Efficient and Compact Neural Networks," demonstrates how attaching auxiliary classifiers to shallow layers can help them learn from deeper layers, leading to increased efficiency and accuracy. If a classification model consists of 100 layers and the features extracted by a relatively shallow layer, such as layer 40, are sufficient to make reasonably accurate predictions, it is likely that the deeper layers (e.g., layers closer to 100) will produce even more confident and accurate classifications. This is because deeper layers typically build on the representations learned by earlier layers, capturing more abstract and high-level features that contribute to stronger predictive performance.

While offline distillation uses a fully trained, high-capacity teacher model, it can often be less effective than online distillation, which uses a teacher model that is still in the process of learning. This might seem counterintuitive at first since the offline teacher has already mastered the task. However, this mastery often leads to overconfidence in predictions.

For instance, a trained language model might consistently predict the token *bright* following the prompt "*The sky is.*" A student trained using offline distillation may simply memorize such mappings, potentially resulting in overfitting and reduced generalization.

On the other hand, online distillation allows the student to learn at the same time with the teacher. Instead of just copying static outputs, the student receives evolving guidance as the teacher improves. This provides the student with the opportunity to learn more patterns from the data, resulting in a richer and more adaptive learning experience.

To better understand the difference between offline and online distillation, consider the following analogy:

- Offline distillation is like attending a one-hour lecture by an expert on an advanced topic. The expert presents the information all at once, assuming a high level of prior knowledge. The students might struggle to absorb or follow the lecture in full.

- Online distillation, by contrast, has both the expert and the student learning a new topic together. As the expert begins to understand the subject, they may make mistakes, ask questions, and refine their understanding, just like the student. This co-learning environment allows the student to grasp the topic more gradually and thoroughly, often gaining insights from the expert's evolving thought process.

Response-Based

Since response-based knowledge distillation is the most widely used form of distillation, we illustrate its mechanism with an example based on HuggingFace.

Figure 3-15 provides an overview of how response-based KD works. In this process, a prompt is first fed into the teacher model, typically an LLM. The teacher generates a response, which is assumed to be high-quality due to the teacher's strong performance. This response is then saved and paired with the original prompt, forming a new synthetic dataset. Once all prompts in the original dataset have been processed in this manner, the role of the teacher concludes.

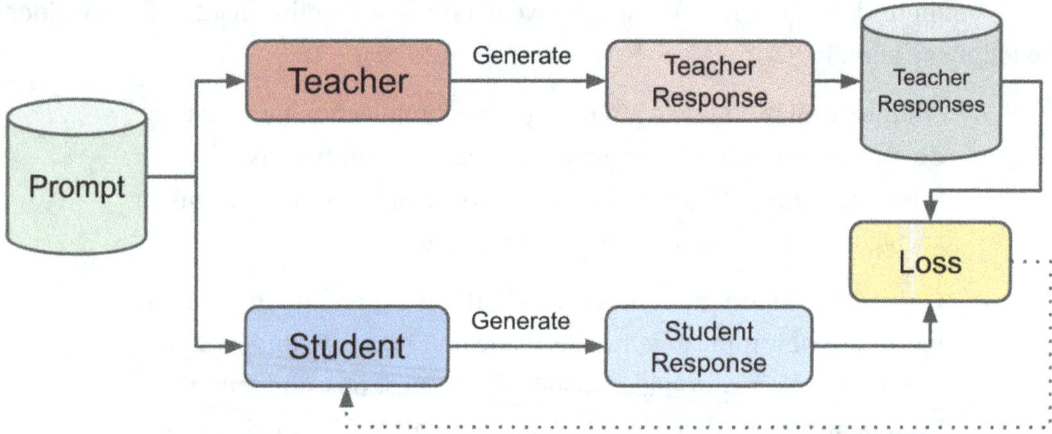

Figure 3-15. *Steps of response-based knowledge distillation. A prompt is fed to both the teacher (a large language model) and the student (a small language model). The teacher's response is treated as a high-quality reference and is used to supervise the training of the student model. The student learns to generate responses that closely match those of the teacher, enabling efficient knowledge transfer*

Next, the student model, which is usually an SLM, is trained using this synthetic dataset. For each prompt, the student generates its own response. The training is supervised, as the student's output is compared to the corresponding teacher-generated response, which serves as the target or ground truth.

A loss function is used to quantify the difference between the student's output and the teacher's response. The training objective is to minimize this loss, effectively teaching the student to approximate the teacher's behavior. In the ideal case, the student learns to generate responses that are nearly identical to those of the teacher, but at a fraction of the computational cost.

This approach is particularly powerful when labeled data is scarce or unavailable, as it depends on the teacher to create high-quality supervision signals for the student. Even some LLMs are trained while a portion of their data is synthetic.

To demonstrate how response-based KD works, let's walk through a simple example using Hugging Face's `transformers` library. In this example, we'll use `GPT2-XL`, a model with 1.5 billion parameters, as the teacher, and `DistilGPT2` as the student, which has just 82 million parameters. While GPT2-XL isn't a large model by today's standards, it still illustrates the concept well. In practice, teacher models often have many billions of parameters.

We'll use the Explain Like I'm 5 (ELI5) dataset, which contains a collection of question-answer pairs. For demonstration purposes and to keep memory usage and training time manageable, we'll use only the first 500 examples.

```python
from datasets import import load_dataset

dataset = load_dataset("sentence-transformers/eli5", split="train[:500]")
dataset = dataset.map(lambda x: {"question": x["question"],
                                 "answer": x["answer"]},
                      remove_columns=dataset.column_names)
```

Here's a sample from the dataset:

```
Question:
Is there a particular shape coffee/tea cup that keeps the beverage
hotter, longer?

Answer:
Spherical. A sphere has the least amount of surface area per volume, and
the surface is where the heat transfer happens. Less surface area means
less heat transfer is possible. Since a sphere is not a practical shape for
a mug, the cylinders we use now is probably the best comprimise.
```

We begin by loading the teacher model using `AutoModelForCausalLM`, which is designed for text generation tasks. Since we're running everything on a CPU, we set the device accordingly. The tokenizer is also loaded using `AutoTokenizer`. These `Auto*` classes automatically infer the correct model/tokenizer architecture based on the passed name.

While we assume the teacher model is well-trained and capable of generating high-quality responses on the ELI5 dataset, this example focuses solely on the mechanics of distillation, not on achieving state-of-the-art performance.

```python
from transformers import AutoTokenizer, AutoModelForCausalLM
import torch

device = torch.device("cpu")
teacher_model = AutoModelForCausalLM.from_pretrained("gpt2-xl").to(device)
teacher_tokenizer = AutoTokenizer.from_pretrained("gpt2-xl")
```

Once the model and its tokenizer are loaded, the next step is to generate responses for the 500 samples using the teacher model. Since this process is repeated for each sample, it's more efficient to define a function and apply it using a mapping approach.

In this example, the generate_teacher_response() function is defined, which takes a single sample from the dataset as input. The function tokenizes the input using the teacher's tokenizer to produce the embedding vector, which is then passed to the model's generate() method to produce the teacher's response. The output is limited to a maximum of 50 tokens.

```python
def generate_teacher_response(example):
    question = example["question"]
    inputs = teacher_tokenizer(question,
                               return_tensors="pt",
                               truncation=True).to(device)
    outputs = teacher_model.generate(**inputs,
                                     max_new_tokens=50,
                                     pad_token_id=teacher_tokenizer.eos_
                                     token_id)
    decoded = teacher_tokenizer.decode(outputs[0], skip_special_
    tokens=True)

    # Remove the question from the decoded output
    teacher_response = decoded[len(question):].strip()

    return {"teacher_output": teacher_response}

dataset_teacher = dataset.map(generate_teacher_response, batched=False)
```

Since the teacher model generates its output as a list of token IDs, the output must be decoded back into readable text. The decoded result includes both the original prompt (or question) and the generated response, concatenated together. However, since we're only interested in the generated response, the prompt is removed from the output, and only the response is returned.

Next, we load the DistilGPT2 model as the student, along with its tokenizer. These are loaded in the same way as the teacher model. However, because DistilGPT2 does not come with a padding token by default, we need to manually add one. This is necessary since all prompts will later be padded to a fixed maximum length of 512 tokens in the code.

```python
from transformers import AutoModelForCausalLM, AutoTokenizer

student_model = AutoModelForCausalLM.from_pretrained("distilgpt2").
to(device)
student_tokenizer = AutoTokenizer.from_pretrained("distilgpt2")

if student_tokenizer.pad_token is None:
    student_tokenizer.pad_token = student_tokenizer.eos_token
```

Since the student model may use a different tokenizer than the teacher, the training samples must be re-tokenized using the student's tokenizer. This is handled by the `tokenize_student()` function, which takes a single dataset sample as input.

For each sample, the prompt is constructed by concatenating the question from the ELI5 dataset with the corresponding response generated by the teacher model. This combined text is then tokenized using the student's tokenizer. The maximum sequence length is set to 512 tokens, ensuring that shorter sequences are padded accordingly.

The tokenizer input, consisting of both the question and the teacher's response, allows the student model to learn how to generate appropriate responses given a prompt. However, during training, we want the student to predict only the response portion. To achieve this, the tokens corresponding to the question are masked using the special label ID -100, which instructs the loss function to ignore them. As a result, the student learns to generate only the response tokens. The `tokenize_student()` function returns a dictionary containing the tokenized and labeled input.

```python
from transformers import DataCollatorForLanguageModeling

def tokenize_student(example):
    full_text = example["question"] + " " + example["teacher_output"]
    tokenized = student_tokenizer(full_text,
                                  truncation=True,
                                  padding="max_length",
                                  max_length=512)

    input_ids = tokenized["input_ids"]
```

```
    question_len = len(student_tokenizer(example["question"],
                                    add_special_tokens=False)
                                    ["input_ids"])

    labels = [-100] * question_len + input_ids[question_len:]
    labels = labels[:512]

    tokenized["labels"] = labels
    return tokenized

tokenized_dataset = dataset_teacher.map(tokenize_student,
                                    remove_columns=dataset_teacher.
                                    column_names)

data_collator = DataCollatorForLanguageModeling(student_tokenizer,
mlm=False)
```

The `tokenize_student()` function is applied to each sample in the training dataset using the `map()` method. This ensures that all samples are tokenized consistently using the student model's tokenizer.

Next, a data collator is created to form batches of training samples. Since the student model is a causal autoregressive model, masked language modeling (MLM) is disabled to avoid masking tokens in the input. The collator is also responsible for padding the sequences in each batch to match the length of the longest sample in that batch. However, padding has already been handled earlier during tokenization.

With the training data prepared, the next step is to define the training arguments and initialize the `Trainer`. The parameter `per_device_train_batch_size` should be adjusted based on the available memory of the CPU or GPU being used. Finally, training is initiated by calling the `train()` method on the `Trainer` instance.

```
from transformers import Trainer, TrainingArguments

training_args = TrainingArguments(
    output_dir="./student-distilled",
    per_device_train_batch_size=8,
    num_train_epochs=10,
    logging_steps=10,
    save_strategy="no",
)
```

```python
trainer = Trainer(
    model=student_model,
    args=training_args,
    train_dataset=tokenized_dataset,
    tokenizer=student_tokenizer,
    data_collator=data_collator,
)

trainer.train()

student_model.save_pretrained("./distilled-student")
student_tokenizer.save_pretrained("./distilled-student")
```

After training is complete, it is useful to compare the responses generated by the teacher and student models on a few sample prompts.

To quickly load the trained student model for text generation, a `pipeline` is created. Since predictions are performed on the CPU, the `device` parameter is set to `-1`. The same question is then passed to both the teacher and the trained student models, and their responses are printed for comparison.

```python
from transformers import pipeline

teacher_generator = pipeline("text-generation",
                             model=teacher_model,
                             tokenizer=teacher_tokenizer,
                             device=-1)

for ex in examples:
    question = ex["question"]
    print("Question:")
    print(question)

    student_output = student_generator(question,
                                       max_new_tokens=50,
                                       do_sample=False)[0]
                                       ["generated_text"]
    student_response = student_output[len(question):].strip()
```

```
teacher_output = teacher_generator(question,
                                   max_new_tokens=50,
                                   do_sample=False)[0]
                                   ["generated_text"]
teacher_response = teacher_output[len(question):].strip()

print("Teacher Response:")
print(teacher_response)
print("Student Response:")
print(student_response)
```

Below is an example question from the dataset:

```
How are police sketch artists able to draw people relatively accurately
based on descriptions?
```

The teacher model responded with the following, which shows a coherent response. Note that the response is truncated due to the 50-token generation limit.

```
The police are trained to be able to identify people based on their
physical appearance. They are trained to be able to identify people based
on their clothing, their hair, their eyes, their skin color, their height,
their weight, their build ...
```

After training, the student model produces a very similar response, with only minor variations:

```
The police are trained to be able to identify people based on their
appearance. They are trained to identify people based on their clothing,
their hair, their eyes, their skin color, their height, their weight, their
build quality, ...
```

Before training, the student model struggled to generate meaningful responses and often produced incomplete or empty outputs. After training through response-based distillation, the student clearly begins to mimic the teacher's behavior more closely.

Here are two additional examples showing near-identical outputs from the student and teacher.

```
Example 1
Question:
```

How can someone just walk away from a home loan?

Teacher Response:
The answer is simple: because they can't afford it. The average home loan in Canada is $1.3 million, according to the Canadian Real Estate Association. That's more than double the average household income of $66,890 ...

Student Response:
The answer is simple: because they can't afford it. The average home loan in Canada is $1.3 million, according to the Canadian Real Estate Association. That's more than double the average household income of $66 ...

Example 2
Question:
Stars and Visibility

Teacher Response:
The Sun is a very bright object, and it is visible to the naked eye from most of the world. The Sun's brightness is measured in terms of its apparent magnitude, which is the apparent brightness of the Sun as seen from Earth. ...

Student Response:
The Sun is a very bright object, and it is visible to the naked eye from most of the world. The Sun's brightness is measured in terms of its apparent magnitude, which is the apparent brightness of the Sun as seen from Earth. The Sun ...

These examples illustrate the effectiveness of response-based KD, a powerful technique for compressing LLMs, which enables a lightweight student model to approximate the behavior of a much larger teacher model. This makes response-based KD a powerful technique for rapid prototyping and deployment of smaller models in environments with limited resources.

However, care must be taken during training to avoid overfitting, as the student may learn to memorize rather than generalize beyond the training examples.

3.5 Training

Fine-tuning a model is not always the optimal choice, especially when sufficient training data is available to start training from scratch. One reason is that the base model may carry inherent biases from its original training data, potentially influencing the fine-tuned model toward certain viewpoints.

Furthermore, using a pre-trained model constrains you to its original tokenizer. If the nature or distribution of your target data differs significantly from that of the base model's training data (e.g., differences in vocabulary, domain-specific terms, or language style), this mismatch can degrade performance or even make the model unsuitable for the task.

For example, a base model may use a vocabulary extracted from English natural language text. While such a model might still function with programming languages, it would not be optimized for them. Similarly, if the tokenizer of the base model is built for English but the target data is in Arabic, the vocabulary would be largely unusable. The same performance limitations occur for other specialized data types, such as chemical formulas, MIDI or musical notation files, mathematical expressions, or DNA sequences. In such cases, many tokens may be treated as unknown, leading to poor performance. Under these circumstances, training a model from scratch is often the best choice.

However, deciding to train a model from scratch comes with significant challenges. First and foremost, the quality of the data directly impacts the quality of the resulting model. Since LLMs require massive amounts of data, the dataset will often be assembled from multiple sources, each with different writing styles, formats, and quality levels. This variability may introduce noise or low-quality samples, and it can be difficult to enforce a consistent tone or style, especially if the target application demands it.

Beyond data preparation, the training process itself is highly complex and resource-intensive, requiring substantial computational power, memory, and storage. During training, techniques such as teacher forcing are often employed to accelerate convergence. In a typical decoder, each generated token depends on the token generated before it, which can slow down training. Teacher forcing mitigates this by feeding the decoder the ground truth token from the training data at each step instead of waiting for the model's own predicted token, enabling faster and more stable training.

The training process involves multiple stages until producing a trustful model:

1. **Pre-Training Self-Supervised Learning (SSL)**: The model predicts the next word in a sequence and compares it to the actual word in the training text. This initial stage equips the model with broad knowledge about language, facts, and patterns within a domain or across multiple fields.

2. **Supervised Fine-Tuning (Supervised Learning or Supervised Fine-Tuning or Instruction Tuning)**: The model is customized for a specific task, such as question answering or text summarization, by explicitly providing it with instructions to follow. For example:

   Help me find the answer to the question "What is the capital of ancient Egypt?" from this text: "In the ancient times, the Egyptian capital is ...."

   This stage tests the model on unseen text, ensuring it can follow instructions rather than simply predict the next token. After instruction tuning, the model evolves from being a general next-token predictor into an assistant that can understand and execute human instructions.

3. **Reward Model Training (Supervised Ranking Training)**: Train a reward model using the human evaluation data to mimic their behavior in evaluating the model responses.

4. **Reinforcement Learning**: Use the trained reward model to train the model. Because humans are the evaluators, it is called Reinforcement Learning from Human Feedback (RLHF).

5. **Post-Training**: This optional step might inject some ethics or safety concerns to, for example, force the model to respect policies of a company, reject answering specific kinds of questions, etc. It might also involve adapting the model into a specific domain.

3.5.1 Self-Supervised Learning (SSL)

In self-supervised learning (SSL), the model treats each next token prediction as a training sample. For example, consider the prompt:

```
The weather today is very good
```

The model is first given only the first token, The, and is then asked to predict the next token:

```
The -> weather -> today -> is -> very -> good
```

At each step, the model's predicted token is compared to the actual next token in the sequence. This comparison produces a loss signal, which is used to update the model's parameters. Because the target output at each step is derived directly from the input text, this process constitutes self-supervised learning.

3.5.2 Supervised Fine-Tuning (SFT)

Supervised fine-tuning (SFT) is typically performed before applying reinforcement learning from human feedback (RLHF). In this stage, a pre-trained model is fine-tuned on a dataset of ideal responses, which may be curated by human labelers. The process is similar to standard supervised learning, where the model is trained to produce outputs that match the provided examples.

SFT applies the same token-by-token prediction strategy used in SSL. At each step, the model's predicted next token is compared with the actual token in the target text, and the resulting loss is used to adjust the model's parameters. Over time, this guides the model to follow specific instructions when generating responses.

Examples of such instructions include:

1. Starting the answer with a greeting message.

2. Ending the answer with a message inviting the user to ask additional questions.

3. Acting as a medical advisor to provide the best possible medical guidance.

4. Generating a concise summary of the input text.

Through this process, SFT adapts the base model to align with desired response styles, formats, or behaviors, preparing it for further refinement through RLHF.

3.5.3 Reward Model

In the reinforcement learning from human feedback (RLHF) process, the fine-tuned model is given a set of questions, and multiple responses are generated for each question. Human evaluators then rank these responses based on their quality. These rankings are used to train a reward model, which is a neural network designed to mimic human preferences in selecting the best answer.

Once the reward model has been trained on sufficient human-labeled data, it can replace human evaluators for future scoring. For any response generated by the LLM, the reward model assigns a reward score, with higher scores given to better responses. Using these scores, a loss function is computed to guide further optimization of the LLM.

For example, let r denote the reward function, A the preferred (high-quality) response, and B a less desirable or incorrect response. If the reward model correctly assigns a higher score to A than to B, the loss will be small.

$$L = -\log\left(\frac{e^{r(A)}}{e^{r(A)} + e^{r(B)}}\right) \tag{14}$$

In one case, if the correct response is assigned a probability of 0.8, the resulting loss might be as low as 0.0969, reflecting a correct ranking. Conversely, if the correct response is assigned a very low probability (e.g., 0.001), the loss could be as high as 3, penalizing the model for ranking responses incorrectly.

Through repeated optimization, the reward model learns to align its scoring with human judgment, enabling the LLM to produce responses that better reflect human preferences without requiring manual evaluation for every example.

3.5.4 Reinforcement Learning (RL)

Once both the supervised fine-tuned model and the reward model are ready, the next step is to apply reinforcement learning (RL) to further align the LLM with desired behaviors. In this phase, the LLM (treated as the *policy model*) generates responses to various prompts, and the reward model evaluates each response, assigning a score that reflects its quality. These scores are then used to update the LLM's parameters in a way that maximizes expected rewards.

One commonly used RL algorithm for this purpose is Proximal Policy Optimization (PPO). PPO is particularly well-suited to training large models because it strikes a balance between improving the policy and preventing overly large updates that could destabilize learning. By iteratively generating responses, receiving rewards, and adjusting parameters, the model becomes increasingly aligned with the reward model's scoring criteria. As a result, the model becomes better aligned with human preferences.

3.6 Evaluation

LLMs are powerful but far from perfect. They can perform exceptionally well in some applications (e.g., classification) while being catastrophically wrong in others. The goal of LLM evaluation is to determine whether the model can be trusted by systematically checking several key dimensions:

1. **Accuracy or Correctness**: Whether the generated response is factually correct or contains mistakes.

2. **Harmfulness or Toxicity**: Ensuring the model does not produce harmful or unsafe outputs, especially in domains like healthcare. In such domains, the model should only provide certain guidance (e.g., prescribing drugs or medical tests) if the inputs clearly justify it. For instance, diagnosing diabetes should only occur when there is certified test evidence, such as an A1C level above a medical threshold.

3. **Relevance**: The response must be relevant to the prompt and its context.

4. **Instruction Following**: The model should follow the instructions in the prompt faithfully.

5. **Completeness or Fulfillment**: The answer should address all the requirements stated in the prompt.

Traditional ML and deep learning evaluation methods don't directly apply to LLMs. In conventional ML, we typically have a labeled dataset from which both the training and test sets are sampled from the same distribution. This allows us to compute standard metrics (e.g., accuracy, F1 score, confusion matrix) against known ground truth.

With LLMs, however, production data distributions often differ from any available evaluation dataset. It's not always clear what the correct metric is to assess the model's accuracy, relevance, etc. For example, how would you objectively evaluate a prompt such as *"What is the best city in the world?"* when there is no universal ground truth answer.

Traditional ML models are usually built for specific tasks such as classification (outputting one category from a fixed set). LLMs, by contrast, can handle a wide variety of tasks, including classification, regression, summarization, creative generation, code writing, image captioning, and more. Even within text generation, there are multiple sub-tasks.

This raises two fundamental challenges:

1.   Which data should be used for each type of task?

2.   What metrics should be applied for each task type?

Some of the solutions are:

1.   Using a public evaluation benchmark

2.   Automatic evaluation

3.   Human evaluation

There are many public benchmarks available to evaluate the model. Their problem is that the dataset might not include data related to your task. But they are good to know where your model stands compared to other models.

1.   **BIG-bench**: A broad benchmark covering many task types.

2.   **HELM (Holistic Evaluation of Language Models)**: Evaluates across multiple dimensions such as accuracy, toxicity, and fairness.

3.   **LM Evaluation Harness**: Similar to BIG-bench, it provides few-shot evaluation over many tasks.

4.   **HumanEvaL**: A code-generation benchmark from OpenAI with 164 Python tasks. If any generated solution passes the tests, the model wins that task. It's useful for code correctness but limited in scope (e.g., no docstring generation or unit-test coverage).

5. **MMLU (Massive Multitask Language Understanding)**: A dataset of multiple-choice questions (MCQs) to test the LLM knowledge against 57 different fields such as computer science, statistics, biology, law, etc.

6. **MuSR (Multistep Soft Reasoning)**: Evaluates the reasoning capabilities of the LLMs through steps requiring multiple steps/turns.

In automatic evaluation, another model is used to assess the target LLM's responses. This approach is faster and more scalable but carries risks, as the evaluator model may lack domain-specific knowledge or introduce biases into the assessment.

Regarding human evaluation, experts create prompts along with ideal, ground-truth responses. These prompts are then fed to the LLM, and its outputs are compared to the ground truth. The comparison can be performed automatically using semantic similarity metrics or manually by having human evaluators rate the responses across multiple dimensions such as accuracy, relevance, and completeness. However, generating high-quality ground-truth responses is both challenging and costly, especially when it requires domain specialists.

For production systems, the best approach is often to build a custom evaluation dataset that reflects the range of real-world cases your application will face. This dataset should evolve over time to cover new scenarios and edge cases.

For example, if your LLM generates mobile device specifications, your evaluation set might initially contain models from major brands. As new devices launch, you would expand the dataset to include them.

If your product has comprehensive documentation, you can automate question-answer pair generation using tools like LangChain's auto-evaluator. It uses an LLM to draft evaluation prompts and answers, which humans then review rather than creating from scratch.

3.6.1 Metrics

The metrics used to evaluate LLMs can be broadly categorized into four groups:

1. **Traditional (Reference-Based)**: Compare the generated responses against predefined reference answers using metrics such as accuracy or BLEU.

2. **Semantic (Reference-Free)**: Evaluate responses based on their meaning or similarity in an embedding space, without requiring exact reference texts.

3. **LLM-Based**: Use another language model as an evaluator to assess the target LLM's outputs.

4. **Human**: Involve human evaluators to rate responses according to dimensions such as accuracy, relevance, and completeness.

Human evaluation is particularly valuable when reference answers are unavailable or when the task scope cannot be precisely defined for automatic evaluation by another LLM.

Traditional Reference-Based

Traditional machine learning (ML) evaluation metrics (e.g., accuracy, precision, and recall) are applicable only when labeled, ground-truth data is available. They are most effective when there is a clearly defined correct answer, making evaluation essentially binary. If the model's output matches the correct answer, it is considered positive; otherwise, it is negative.

When evaluating text generated by an LLM, reference matching techniques are often employed. One example is the Levenshtein distance, which measures the minimum number of single-character edits required to transform one string into another. Another widely used approach is n-gram overlap, which counts how many consecutive word sequences (n-grams) in the generated output also appear in the reference text. Within n-gram–based methods, BLEU emphasizes precision and is commonly applied to tasks such as machine translation and text rephrasing, while ROUGE focuses on recall and is frequently used for summarization, assessing how much of the reference content is preserved in the generated output.

Semantic

Reference-based metrics depend on a reference answer to calculate the metric. Since it is almost impossible to create reference answers for all the possible prompts received by the LLM, semantic-based metrics are more flexible, as they can be used with or without ground truth. When ground truth is available, it measures the semantic similarity between generated and reference texts. Without ground truth, it measures the similarity between the prompt and the generated response to ensure topical relevance.

BERTScore and MoverScore are semantic similarity metrics that work by converting text into high-dimensional feature vectors representing its meaning. The similarity between two texts is then computed using a distance measure such as cosine similarity, producing a score between 0 and 1, where 1 indicates semantic equivalence and 0 indicates complete dissimilarity.

While these metrics are valuable for certain tasks, they are not always ideal for evaluating LLM-generated responses, as the feature vectors do not necessarily have a direct correlation with the reference texts. Two feature vectors with different semantics can have similar results using semantic-based metrics. However, they are highly effective in training and evaluating traditional classification models.

LLM-Based

LLM-as-a-Judge evaluation works with or without ground-truth data and involves using an LLM to assess the quality of responses generated by another model. In its simplest form, the judging LLM can be asked to choose the better response from a set, such as *"What is the best answer? {A} {B}."*

This approach can be problematic when the definition of the best answer is not clear. Without explicit criteria, the judge may apply its own subjective preferences, as some models tend to favor longer answers regardless of quality, while others may exhibit positional bias toward the first option. To reduce such biases and align evaluation with the intended standards, the judging LLM may require fine-tuning or clear guidance.

When assessing free-text generation, it is often necessary to provide context and structured evaluation criteria. For example, a scoring template can be given, and the LLM asked to rate a response from 1 to 5 based on adherence to specific guidelines. LLM-as-a-Judge can also operate on single responses, classifying them (e.g., "Is this about food or drinks?") or assigning a rating (e.g., "Rate from 1 to 5 for relevance to this document").

Supplying carefully crafted reference examples can further improve consistency, helping the model decide whether a response meets defined standards. The key to effective use of this approach is to be as specific as possible in the evaluation criteria. Rather than asking if a response is simply good or bad, the prompt should clearly define what constitutes a good or bad response.

Human

Since LLMs may exhibit biases, it is generally advisable to have human evaluators oversee the model's judgments to ensure some factors that cannot be easily evaluated automatically, such as fairness. For instance, RLHF evaluators score model responses on a scale from 1 to 5, with 5 representing the highest quality.

However, human evaluation also presents challenges. A single evaluator may not be consistent when rating similar answers, making it essential to establish a clear and well-defined set of evaluation criteria. Consistency can be improved by involving multiple evaluators and averaging their scores, which helps reduce individual bias and variability. This approach is both complex and costly, as it requires identifying many qualified evaluators with relevant domain expertise and monitoring their work to ensure adherence to the established standards. This is also costly.

3.7 Evolution of Popular Models

Monitoring the development of LLMs by major vendors in the generative AI industry provides valuable insight into how competitive and fast-growing the models' development is. In this section, we place particular emphasis on OpenAI and DeepSeek. The timeline of the model releases is summarized in Figure 3-16.

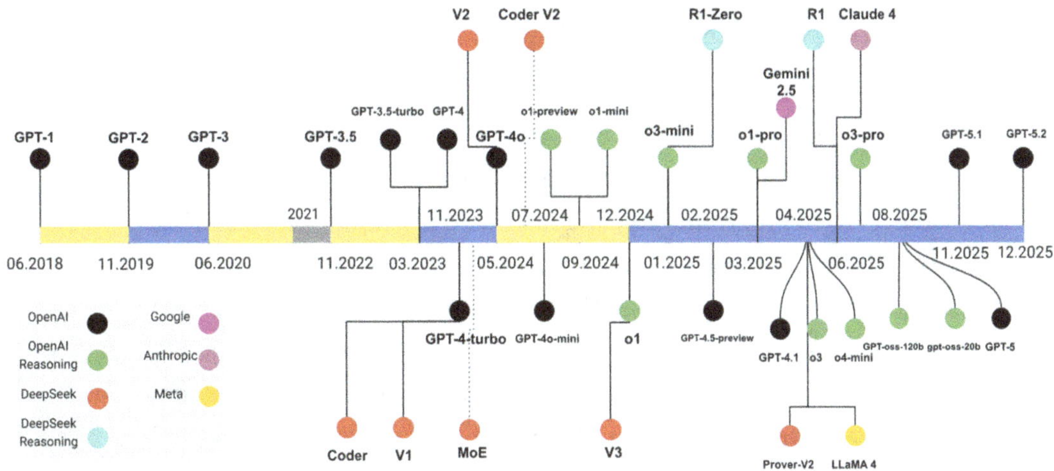

Figure 3-16. *Timeline showing the release history of OpenAI and DeepSeek models in addition to the latest models from Google, Meta, and Anthropic. This is not a comprehensive list of all models, and it just highlights the key milestones in the generative AI model development*

OpenAI was among the first research organizations to successfully develop fully functional generative AI models, establishing itself as a leading innovator in the field. DeepSeek introduced a major shift in the industry by releasing high-performance cost-efficient models as open source, enabling broader access for researchers and practitioners.

The section also highlights the latest models, up to the time of writing the book, developed by Google, Anthropic, and Meta.

Even though we are listing many models developed by OpenAI and DeepSeek, this is not the full comprehensive list of all models developed by them.

3.7.1 DeepSeek

DeepSeek is a family of large language models that compete with industry-leading systems such as OpenAI's GPT, Google's Gemini, and Meta's LLaMA. It is recognized for delivering strong performance while significantly reducing computational and deployment costs. The most advanced model in the series to date, DeepSeek-R1, is a reasoning-focused model that not only generates answers but also allows users to inspect its intermediate reasoning steps, providing greater transparency into the decision-making process.

Out of the full list of LLMs developed by DeepSeek, the following is a selected list of some of the most important releases:

1. DeepSeek-Coder (November 2023) was released in multiple model sizes, ranging from 1 billion to 33 billion parameters. It was trained on 2 trillion tokens, with approximately 87% of the data dedicated to programming languages and the remainder covering English and Chinese natural language content. It supports 86 programming languages. The release marked a significant milestone, drawing attention to the feasibility of developing an open-source model that could approach GPT-level performance while substantially reducing deployment costs.

2. DeepSeek-LLM (V1) (November 2023) is an open-source, 67-billion-parameter model and the first release in the DeepSeek-LLM series. Trained on 2 trillion English and Chinese tokens, it laid the groundwork for all subsequent DeepSeek models.

Alongside the flagship model, a 7-billion-parameter base version was also released to support researchers and facilitate experimentation.

3. DeepSeek-MoE (January 2024) reintroduced the *mixture-of-experts* (MoE) paradigm, a classic machine learning technique adapted for modern large-scale models. Rather than processing every input through a single, full-size neural network, MoE architectures employ N specialized expert networks, each representing a fraction of the overall model size. For a given input, only a small subset of these experts is activated, enabling the model to achieve high performance while significantly reducing the number of parameters engaged during inference. Different model sizes are created with 2B, 16B, and 145B parameters. These models achieve comparable performance with other models (e.g., LLaMA 7B and DeepSeek 67B) while only activating a fraction of the parameters.

4. DeepSeek-V2 (May 2024) is a 236-billion-parameter model in which only 21 billion parameters are activated for each token. The model uses two key features: Multi-head Latent Attention (MLA), which compresses the key–value matrices into latent vectors to reduce memory usage, and MoE, which applies sparse computation to lower training costs while maintaining high performance.

5. DeepSeek-Coder-V2 (June 2024): This is an improvement of a checkpoint of DeepSeek-V2 to use the MoE technique to improve the performance of DeepSeek-Coder. It comes in 16B and 236B parameters. Beyond coding tasks, it also handles mathematical and general language tasks. It expands its support to 338 programming languages, compared to just 86 for DeepSeek-Coder.

6. DeepSeek-V3: December 2024. This is an extension of the DeepSeek-V2 model, which is deeper with 671B parameters but only 37B parameters activated for each token. After being trained on 14.8 trillion high-quality tokens, supervised fine-tuning (SFT)

and reinforcement learning (RL) are applied to further improve the model's performance. On top of DeepSeek-V2, V3 applies an auxiliary-loss-free strategy and a multi-token prediction training objective.

7. DeepSeek-Prover-V2 671B (April 2025) is a large-scale model specialized in advanced mathematical reasoning for formal theorem proving. Designed for use with the *Lean 4* proof assistant, it uses the reasoning capabilities of DeepSeek-V3 to decompose complex proofs into a structured sequence of smaller subtasks.

The reasoning-focused models developed by DeepSeek are:

1. DeepSeek-R1-Zero (January 2025) marked the first generation of DeepSeek's reasoning models, built on the DeepSeek-V3-Base architecture. The "R" in R1 stands for *reasoning*. Unlike typical instruction-tuned models, R1-Zero was trained exclusively through reinforcement learning (RL) without an initial supervised fine-tuning (SFT) stage. While this approach demonstrated promising reasoning capabilities, it also introduced limitations such as repetitive outputs and unintended language mixing within the same response.

2. DeepSeek-R1 (January 2025) extended and refined the R1-Zero approach through a multi-stage training pipeline. The process began with a *cold-start* dataset with thousands of high-quality examples featuring clean chain-of-thought reasoning used to fine-tune the DeepSeek-V3-Base model prior to RL. This solved the primary issues observed in R1-Zero and improved reasoning performance. Next, a reinforcement learning stage was applied. Supervised fine-tuning (SFT) data was then created via *rejection sampling*, in which multiple model outputs were evaluated and low-quality ones discarded. This curated dataset, combined with additional supervised data generated by DeepSeek-V3, was used for further fine-tuning. A final RL stage was applied to consolidate these improvements, resulting in a model with more coherent, accurate, and contextually consistent reasoning capabilities.

The reasoning models in the DeepSeek-R1 series were developed exclusively through RL, without incorporating any supervised fine-tuning data. Training was conducted using the Group Relative Policy Optimization (GRPO) algorithm, a reinforcement learning method tailored for optimizing reasoning performance. Building on the reasoning data produced by DeepSeek-R1, six distilled models were created by fine-tuning smaller-scale versions of the Qwen and LLaMA architectures, enabling more efficient deployment while retaining much of the original model's reasoning capability.

3.7.2 OpenAI

OpenAI was the first research company to introduce a generative language model with the release of GPT-1. In November 2022, it launched ChatGPT, the first widely adopted AI chatbot, which quickly made OpenAI a household name in artificial intelligence. Today, it remains one of the most competitive vendors in the generative AI industry, consistently releasing models with industry-leading capabilities.

OpenAI has released numerous models over the years, but in this section, we will cover only a selected subset. For a comprehensive and up-to-date list of all models developed by OpenAI, refer to the official documentation at `https://platform.openai.com/docs/models`.

1. GPT-1 (June 2018) was the first generative pre-trained transformer (GPT) released by OpenAI, containing 117 million parameters. It was trained on the *BooksCorpus* dataset, which consists of approximately 7,000 unpublished books. While its performance in text generation was limited, GPT-1 represented a pivotal milestone, laying the groundwork for the rapid advancements that followed in the GPT model series. `https://openai.com/index/language-unsupervised`

2. GPT-2 (November 2019) was developed as the successor to GPT-1, featuring a deeper architecture with 1.5 billion parameters. Although its initial release was planned for February 2019, OpenAI withheld the full model for several months due to concerns over potential misuse. GPT-2 was trained on a 40 GB dataset comprising content from more than 8 million web pages. It demonstrated a substantial performance improvement over

GPT-1 in generating coherent text, but its outputs were still prone to factual inaccuracies and inconsistencies, limiting its reliability for critical applications. `https://openai.com/index/gpt-2-1-5b-release`

`https://huggingface.co/transformers/v2.11.0/model_doc/gpt2.html`

3. GPT-3 (Davinci) (June 2020) was the first model in the GPT series to deliver consistently realistic and contextually accurate responses. It featured a substantially more complex architecture with 175 billion parameters and was trained on approximately 300 billion tokens of text, drawn primarily from diverse Internet sources. This breadth of training data enhanced the model's understanding of language, improved its ability to engage with a wide range of topics, and enabled it to generate human-like text outputs. In addition to the flagship Davinci variant, the GPT-3 family included the smaller legacy models Ada, Babbage, and Curie, each offering different trade-offs between speed and capability. `https://openai.com/index/language-models-are-few-shot-learners`

4. GPT-3.5 (November 2022) was developed as a fine-tuned successor to GPT-3. The development of GPT-3.5 began in early 2022 with the goal of producing a more robust and reliable system capable of practical, real-world use. While GPT-3 was trained purely through unsupervised learning without direct human involvement, GPT-3.5 incorporated supervised fine-tuning (SFT) and reinforcement learning from human feedback (RLHF). These techniques enabled the model to generate outputs that are better aligned with human preferences, reduced hallucinations, and embedded considerations for safety and privacy into its responses. This fine-tuning laid the foundation for ChatGPT, a conversational AI built on the GPT-3.5 architecture, designed to follow user instructions and engage in natural, contextually aware dialogue.

5. GPT-3.5-Turbo (March 2023) is a cost-optimized variant of
    GPT-3.5 designed to deliver faster responses, particularly for
    simpler tasks. It is specifically optimized for conversational
    applications, making it a more efficient and economical choice
    for powering ChatGPT and other interactive systems, while
    maintaining much of the capability of the original GPT-3.5 model.

6. GPT-4 (March 2023) was the first multimodal model in the GPT
    series. It can accept both text and image inputs while producing
    text outputs. The initial release had a context window of 8,000
    tokens, and a later version increased this to 128,000 tokens.
    This is a significant improvement over the 4,000-token limit of
    GPT-3.5. In casual text conversations, GPT-4 performed similarly
    to GPT-3.5. Its strengths became more evident in complex tasks,
    where it showed greater creativity, improved reasoning, and better
    adherence to nuanced instructions. Despite these improvements,
    GPT-4 still had limitations in certain scenarios and sometimes
    underperformed compared to human experts in benchmark
    evaluations. `https://openai.com/index/gpt-4-research`

7. GPT-4 Turbo (November 2023) is an enhanced version of GPT-4
    that supports a context window of 128,000 tokens. It is designed
    to be more cost-effective than the original GPT-4 and includes a
    more recent knowledge cutoff, allowing it to incorporate a broader
    range of up-to-date information. `https://platform.openai.com/`
    `docs/models/gpt-4-turbo`

8. GPT-4o (May 2024) is a more powerful multimodal model capable
    of processing text, images, audio, and video as inputs to generate
    text-based responses. The letter *o* stands for *omni*, reflecting the
    model's ability to handle multiple types of data. Upon its release,
    OpenAI recommended that users transition from GPT-4 Turbo to
    GPT-4o for improved performance and broader input capabilities.
    `https://openai.com/index/hello-gpt-4o`

9. GPT-4o-mini (July 2024) is a smaller and more cost-efficient version of OpenAI's GPT-4o model. It is designed to deliver strong performance while reducing computational requirements and operational costs. OpenAI recommended that users of GPT-3.5-Turbo migrate to GPT-4o-mini for improved efficiency and capability. `https://openai.com/index/gpt-4o-mini-advancing-cost-efficient-intelligence`

10. GPT-4.5 Research Preview (February 2025) is a pre-release version of OpenAI's largest model, introduced as a research-only offering rather than a final product. It focuses on scaling unsupervised learning to enhance the model's ability to recognize complex patterns without relying on explicit reasoning. The model integrates expanded unsupervised learning, scaled reasoning techniques, and knowledge distilled from smaller models to achieve higher performance across a wide range of tasks. `https://openai.com/index/introducing-gpt-4-5`

    1. The model gains deeper language knowledge through scaled pre-training on large datasets with an optimized architecture, and post-training with SFT and RLHF makes its responses more natural and aligned with user instructions.

    2. GPT-4.5 offers more natural reasoning, producing more accurate answers to complex questions with a lower hallucination rate than GPT-4o and better performance on various reasoning tasks. It still lags behind specialized reasoning models such as OpenAI's o3-mini, which outperforms it on reasoning benchmarks. Optimized for faster responses in production, GPT-4.5 is the last model in the series without an integrated chain-of-thought (CoT), which OpenAI plans to include in future releases.

    3. It is also trained on data generated by smaller, specialized models fine-tuned for tasks such as improving accuracy, focusing on subtle details, and making precise inferences. For example, it may infer that a user is happy and excited if they are planning a trip abroad. All such data is reviewed by humans to ensure quality, enabling GPT-4.5 to adopt the nuanced understanding exhibited by these smaller models.

11. GPT-4.1 (April 2025) introduced major improvements in coding
     ability, instruction following, and context length, supporting
     up to 1 million tokens. It outperforms GPT-4o, particularly in
     programming-related tasks. The release included two smaller
     variants, GPT-4.1 mini and GPT-4.1 nano, designed for efficiency
     at reduced computational cost. With the launch of GPT-4.1,
     OpenAI announced the deprecation of the GPT-4.5 Research
     Preview, which will be discontinued by mid-July 2025 due
     to GPT-4.1's superior performance. `https://openai.com/index/gpt-4-1`

12. GPT-5 (August 2025) is a unified system that employs a routing
     mechanism to determine whether a query can be answered
     directly or requires deep reasoning. It rapidly evaluates the
     complexity of the request and selects the most suitable processing
     path, enabling both efficient handling of simple tasks and more
     thorough reasoning for complex problems. As advertised by
     OpenAI, GPT-5 is claimed to achieve a level of reasoning and
     problem-solving comparable to that of a PhD student. `https://openai.com/index/introducing-gpt-5`

13. GPT-5.1 (November 2025) introduced adaptive reasoning, which
     allows the model to automatically decide when deeper thinking
     is needed before responding. While this behavior is an intended
     feature, some early user feedback reported uneven response
     quality and an overuse of reasoning in certain scenarios shortly
     after release, but these issues were not formally documented by
     OpenAI. `https://openai.com/index/gpt-5-1`

14. GPT-5.2 (December 2025) emphasizes practical deployment to
     maximize impact in real-world work flows. It delivers stronger
     performance in creating and editing spreadsheets, generating
     higher-quality presentations end-to-end, improved image
     understanding, and more reliable integration with external
     tools, which enable more capable agentic AI systems. OpenAI
     reports that GPT-5.2 outperforms GPT-5.1 in reasoning quality
     across multiple benchmarks. `https://openai.com/index/introducing-gpt-5-2`

Away from the LLMs developed by OpenAI, they also released reasoning models including:

1. o1-preview (September 2024) was an experimental reasoning model released for early testing and research feedback. It served as a precursor to the production o1 model. `https://openai.com/index/learning-to-reason-with-llms`

2. o1-mini (September 2024) is a cost-efficient reasoning model optimized for speed and practicality. `https://openai.com/index/openai-o1-mini-advancing-cost-efficient-reasoning`

3. o1 (December 5, 2024) is OpenAI's first officially released reasoning model, graduating from the September 2024 o1-preview. It "thinks" before answering and lets developers control the reasoning budget to manage cost and latency. It supports both text and image inputs, enabling users to submit visual content for reasoning. `https://openai.com/index/o1-and-new-tools-for-developers`

4. o3-mini (January 2025) is a cost-efficient reasoning model optimized for STEM tasks, especially math and coding. `https://openai.com/index/openai-o3-mini`

5. o1-pro (March 2025) is a higher-compute variant that thinks longer, improving reasoning reliability on complex tasks.

6. o3 (April 2025) supports multimodal reasoning and excels at math, coding, and visual tasks. `https://openai.com/index/introducing-o3-and-o4-mini`

7. o4-mini (April 2025) is a fast, cost-effective reasoning model that achieves strong performance for its size. `https://openai.com/index/introducing-o3-and-o4-mini`

8. o3-pro (June 2025) is a higher-compute model designed for advanced thinking to handle ambiguous and complex problems. `https://help.openai.com/en/articles/9624314-model-release-notes`

9. gpt-oss-120b and gpt-oss-20b (August 2025) are open-weight reasoning models released by OpenAI that process text only. They are designed to allow users to fine-tune and customize the models on their own data, enabling domain-specific adaptation while retaining strong general reasoning capabilities. Although only the model weights are available, this release marks a significant step for OpenAI, which historically has shared little information or assets related to its models. `https://openai.com/index/introducing-gpt-oss`

From the release pattern, OpenAI typically launches a mini version of a reasoning model first, followed by the full model and then a pro variant. Based on this trend, the release of o4 and o4-pro is likely to follow, potentially alongside the debut of an o5-mini model.

3.7.3 Other Vendors

Many vendors develop generative AI models, including Google, which released Gemini 2.5 on March 25, 2025, as an improvement over Gemini 2.0. This model supports native multimodality and offers a long context window of 1 million tokens. To accommodate varying performance and cost requirements, Google provides three versions: Pro, Flash, and Flash-Lite, allowing users to select the option best suited to their needs.

Given that an average token contains about 4 characters, and the average English word length is approximately 4.7–5 characters, one token corresponds to roughly 0.75 words. A context window of 1 million tokens therefore represents about 750,000 words, which may be an underestimate. With an estimated 550 words per single-spaced A4 page in 12-point font, this equals roughly 1,300 pages of text. This is an enormous amount of information to supply in a single prompt.

Meta released its LLaMA (Large Language Model Meta AI) 4 family of multimodal models (`https://ai.meta.com/blog/llama-4-multimodal-intelligence`) in April 2025, following a year after the launch of LLaMA 3 in April 2024. The LLaMA 4 lineup includes:

1. **LLaMA 4 Behemoth**: A model with 288 billion activated parameters serving as the teacher model for the smaller variants.

2. **LLaMA 4 Maverick**: A model with 17 billion activated parameters supporting a 1-million-token context window.

3.   **LLaMA 4 Scout**: A model with 17 billion activated parameters with an extended context window of 10 million tokens.

Behemoth provides the foundation for Maverick and Scout through knowledge distillation, allowing smaller models to retain strong performance while being more efficient in deployment.

Following the release of Claude 4 in May 2025, Claude Opus 4.1 is released on August 5, 2025, an upgrade with targeted improvements in agentic tasks, real-world coding, and reasoning. Anthropic positions it as their most capable model to date, a drop-in replacement that handles complex, multi-step problems with greater precision. In its announcement (`https://www.anthropic.com/news/claude-opus-4-1`), Anthropic indicates state-of-the-art or superior results on coding tasks versus Google's Gemini 2.5 Pro and OpenAI's o3.

Following the release of Google's Gemini 2.5, OpenAI introduced the GPT-4.5-Preview model on April 13, 2025, less than a month later, matching Gemini's extended context size. Soon after, OpenAI deprecated GPT-4.5-Preview with the launch of GPT-4.1, a lower-version model that, nevertheless, outperformed its predecessor.

A similar pattern emerged when OpenAI released GPT-5 on August 7, 2025, just two days after Anthropic unveiled Claude 4.1 on August 5, 2025, which claimed superior reasoning performance over OpenAI's models. In its announcement, OpenAI emphasized GPT-5's reasoning capabilities and its ability to route queries between quick and deep thinking modes.

This sequence of events suggests that OpenAI is determined to maintain dominance in the generative AI market, even if it means releasing interim or rapidly developed models to ensure it remains a visible and competitive force.

There are many other vendors developing generative AI models, including:

- Amazon with its Titan series, offering models for text generation, search, and embeddings through the Amazon Bedrock platform in AWS (Amazon Web Services). AWS offers a lot of helpful tools to build and deploy generative AI and agentic AI systems, including Bedrock, SageMaker, and Q.

- IBM with the Granite family, designed for enterprise applications with a focus on data privacy, security, and explainability.

- Cohere, with its Command models, optimized for instruction following and retrieval-augmented generation (RAG) in business use cases.

- Mistral AI with its Mistral series, known for high-performance open-weight models.

- xAI offers the Grok series of language models, which are integrated into the X platform (formerly known as Twitter) to provide conversational AI capabilities directly within the social media experience. Users can tag Grok in their posts to receive AI-generated responses, allowing them to gain insights and information without leaving the platform.

CHAPTER 4

Model Customization

Training a large language model is an extremely resource-intensive process due to the need to adapt a massive network architecture over billions or even trillions of tokens. This requires substantial computational power, specialized hardware, and significant financial investment.

An LLM is trained on data that, at best, reflects knowledge up to the point when training happens. Once training is complete, the model's knowledge is fixed within the boundaries of that dataset. For example, a model trained only on data up to 2018 would have no awareness of the COVID-19 pandemic.

Given the enormous volume of new information generated daily, fully retraining an LLM at regular intervals to incorporate updated knowledge is impractical. Instead, more cost-efficient strategies are used to embed new knowledge into existing models without repeating the full training process.

This chapter focuses on fine-tuning the model to customize it for new tasks. It starts with full fine-tuning, which works by adjusting all the base model's weights. Then the partial fine-tuning is discussed using some parameter-efficient fine-tuning techniques.

4.1 Model Customization Techniques

Large language models are typically trained on massive datasets to develop broad language understanding and general reasoning capabilities. However, in many real-world applications, we require these models to perform well on specific tasks, follow certain formats, or align with domain-specific knowledge. Instead of training a new model from scratch, which is both time-consuming and expensive, we can customize an existing LLM using various adaptation strategies.

Generally, there are four common approaches for customizing an already trained LLM:

1. **Prompt Engineering**: Designing effective prompts that guide the model toward producing the desired output without changing its internal parameters.

2. **Fine-Tuning**: Adjusting the model's parameters using additional task-specific data to improve performance on a target application.

3. **Continued Pre-Training**: Extending the model's training with new domain-specific or task-specific data to enhance its knowledge base.

4. **Retrieval-Augmented Generation (RAG)**: Combining the model with an external knowledge source, allowing it to retrieve relevant information before generating a response to enrich its context for producing higher-quality responses.

This chapter will focus on the first two approaches (prompt engineering and fine-tuning), exploring how they can be applied to adapt LLMs for specialized tasks efficiently. RAG will be discussed in detail in Chapter 5.

4.2 Fine-Tuning

Building a high-performing large-scale model from the ground up (training from scratch) is increasingly out of reach for the average organization. To achieve state-of-the-art performance, a model must be exposed to trillions of tokens of data to learn basic language structures, grammar, and world facts. This requires months of continuous processing on massive GPU clusters, costing millions of dollars.

Furthermore, training from scratch requires an enormous, high-quality dataset that most specialized domains simply do not possess. Fine-tuning, by contrast, allows us to take a model that already understands the world and simply teach it the nuances of a specific task, turning an impossible engineering feat into a manageable optimization problem.

Transfer learning refers to the process of using a pre-trained model (i.e., base or foundation model) that has been trained on one task, domain, or dataset, and applying it to a new task. Usually, this base model is trained using huge amounts of data on expensive machines that are not affordable by most users.

The base model may include billions or trillions of parameters (i.e., weights) reflecting the knowledge it extracted from the data. It is almost impractical for everyone to train such a huge model. Transfer learning allows the reuse of learned weights or knowledge in the base model and just adapting the model to the new task or data, as illustrated in Figure 4-1. This significantly reduces the need for large amounts of data or computational resources for the new task.

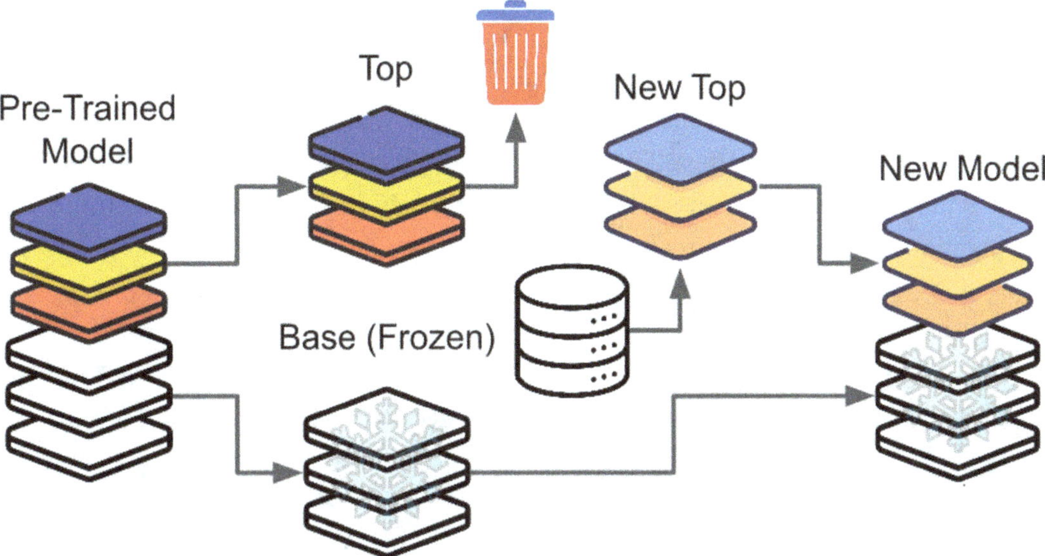

Figure 4-1. *Transfer learning uses the rich knowledge embedded in a pre-trained base model to tackle a new, often smaller or more specialized task. The base model is typically frozen, and only the task-specific top layers are trained*

This approach is particularly common in computer vision, where collecting millions of labeled images across dozens or even hundreds of classes and training models on such large datasets poses significant challenges in terms of time, cost, and computational resources.

Training LLMs from scratch presents significant challenges, primarily due to the high cost associated with collecting and curating the massive datasets required. While building LMs is relatively easy, training LLMs from scratch still remains a very complex task.

In a typical transfer learning pipeline, the base model's parameters are frozen, meaning they are not updated during training on the new task. Instead, new layers are added on top of the model to adapt it to the specific requirements of the target task.

To illustrate, consider a model trained to classify user movie text reviews into one of ten categories from a given dataset. Suppose we want to use this model to perform sentiment analysis, classifying user product text reviews into five sentiment categories from another dataset. Transfer learning would involve removing the original output layers (designed for ten classes) and replacing them with new layers suitable for five-class classification. The base model's weights remain unchanged, and only the newly added layers are trained on the sentiment dataset.

However, in some cases, training only the new top layers may not yield satisfactory performance. Fine-tuning addresses this limitation by unfreezing some (or all) of the base model's parameters and allowing them to be updated during training on the target dataset, as illustrated in Figure 4-2. This additional step enables the model to better adapt to the nuances of the new task, often resulting in improved accuracy but more computational time and resources.

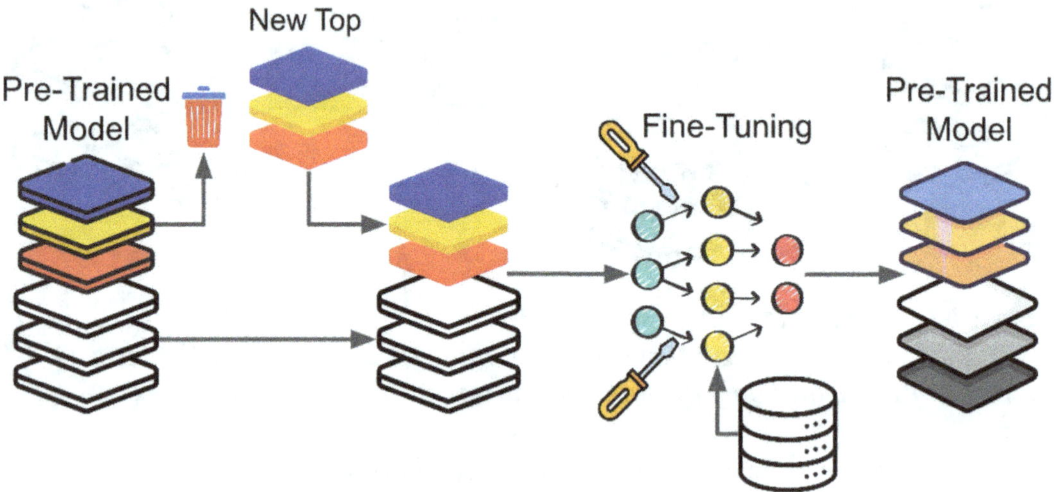

Figure 4-2. *Fine-tuning involves updating all or selected layers of a pre-trained model, in addition to attaching and training task-specific output layers designed for the target problem*

Although the terms of fine-tuning and transfer learning are sometimes used interchangeably, there is a distinct difference between them. Transfer learning refers to using a base model, trained on one dataset (domain or task), and adapting it to a new dataset. When all the base model's weights are frozen, and only the newly added top layer's weights are trained, this is not called fine-tuning. On the other hand, fine-tuning occurs when the model's weights, already trained on a specific dataset, are updated to improve performance on either the same or a different dataset.

Fine-tuning involves updating already learned weights.

Fine-tuning can be categorized based on the scope of parameter updates:

1. **Full Fine-Tuning**: All model parameters (weights) are updated during training, allowing the model to adapt completely to the new task or dataset.

2. **Partial Fine-Tuning**: Only a selected subset of parameters is updated, while the rest of the model remains frozen. This approach reduces computational cost and helps preserve previously learned knowledge.

We will do experiments for each of these fine-tuning approaches using the Hugging Face `transformers` library, providing practical examples and implementation details for both full and partial fine-tuning.

4.2.1 Sequence Classification Fine-Tuning

In this example, we will use an SMS dataset to build a binary classifier that predicts whether a message is **spam** or **not spam**. Before proceeding, ensure that the following Hugging Face libraries are installed:

1. `transformers`

2. `datasets`

3. `evaluate`

Depending on your environment, you may also need additional dependencies such as PyTorch, accelerate, and scikit-learn (required by the `evaluate` library).

We will work with the `ucirvine/sms_spam` dataset, available on the Hugging Face Hub (`https://huggingface.co/datasets/ucirvine/sms_spam`). This dataset contains labeled SMS messages for binary classification, with the following label definitions:

1. 0: Ham (non-spam message)

2. 1: Spam

To load the dataset, we use the following code:

```
from datasets import load_dataset
dataset = load_dataset("ucirvine/sms_spam")
```

As usual with the `datasets` library, the dataset is returned as a `DatasetDict` object. It contains 5,574 samples for training, with two attributes:

1. `sms`: The SMS text to be classified as either ham or spam.

2. `label`: The label for each SMS (0 for ham and 1 for spam).

Here's a representation of the returned `DatasetDict`:

```
DatasetDict({
    train: Dataset({
        features: ['sms', 'label'],
        num_rows: 5574
    })
})
```

In this example, we will use a compact variant of the BERT model, `prajjwal1/bert-tiny`, which is well-suited for demonstration purposes due to its small size and fast training time.

```
model_name = 'prajjwal1/bert-tiny'
```

Next, we load the tokenizer using the `AutoTokenizer` class. This class automatically selects the appropriate tokenizer based on the specified model. Like other *auto* classes in the Hugging Face ecosystem, you can also instantiate a model-specific tokenizer directly if you know it in advance. For example, since our model is based on BERT, we could use the `BertTokenizer` class instead.

```
from transformers import AutoTokenizer
tokenizer = AutoTokenizer.from_pretrained(pretrained_model_name_or_
path=model_name)
```

To tokenize all the samples in the dataset, we define a function named `tokenize_function()` that applies the tokenizer to a batch of samples. We then use the `map()` method to process the entire dataset in batches. This function applies padding to ensure that all sequences have the same length. In this example, we set the `max_length` parameter to 186, specifying the maximum number of tokens allowed per sequence.

```
def tokenize_function(examples):
    return tokenizer(
        text=examples["sms"],
```

```
        padding="max_length",
        max_length=186,
        truncation=True
    )

tokenized_datasets = dataset.map(tokenize_function, batched=True)
```

The `AutoModelForSequenceClassification` class is used to load a pre-trained model for sequence classification tasks. This class automatically selects the appropriate model architecture based on the specified model name. If you know the exact model you wish to load, you can directly use the corresponding model class (e.g., `BertForSequenceClassification` for BERT).

Since each classification problem may differ in the number of classes, it is necessary to configure the model according to the specific classification task. There are several ways to achieve this configuration:

One way is to pass an instance of the `PretrainedConfig` class to the `config` parameter, specifying all necessary configuration parameters. Examples of configuration parameters include:

- `id2label`: A dictionary that maps class IDs to their corresponding labels.

- `label2id`: A dictionary that maps labels to their corresponding class IDs.

- `num_labels`: Defines the number of neurons in the final layer, corresponding to the number of classes in the classification task.

If you know the specific model you are using, you can directly use its corresponding configuration class. For example, `BertConfig` for the BERT model.

Alternatively, you can pass the configuration parameters directly as keyword arguments to the `from_pretrained()` method of the `AutoModelForSequenceClassification` class.

```
id2label = {0: "ham",
            1: "spam"}
label2id = {v: k for k, v in id2label.items()}

model_name = 'prajjwal1/bert-tiny'
```

```
from transformers import AutoModelForSequenceClassification
model = AutoModelForSequenceClassification.from_pretrained(pretrained_
model_name_or_path=model_name,

                                        num_labels=2,
                                        id2label=id2label,
                                        label2id=label2id)
```

The `num_labels` configuration parameter is passed as a keyword argument to specify the number of classes in the classification task. In this case, it is set to 2 to indicate a binary classification problem. This parameter is then used when instantiating an object of the `PretrainedConfig` class.

The used model `prajjwal1/bert-tiny` only contains 4,388,484 parameters, as indicated by the `model.num_parameters()` method. Since it is based on the BERT architecture, the object returned by the `AutoModelForSequenceClassification` class is an instance of the `BertForSequenceClassification` class.

The `id2label` and `label2id` parameters are particularly useful for automatically converting between class labels and their corresponding IDs, especially when using a pipeline for inference. This ensures that the model outputs can be easily interpreted in terms of human-readable labels.

With the dataset, tokenizer, and model prepared, the next step is to configure the training hyperparameters via Hugging Face's `TrainingArguments`. Key settings include learning rate, number of epochs, weight decay, evaluation strategies, checkpointing/saving policy, etc.

```
from transformers import TrainingArguments
training_args = TrainingArguments(output_dir="sms_fine_tuning",
                            logging_dir='logs',
                            learning_rate=0.00001,
                            num_train_epochs=10,
                            weight_decay=0.01,
                            eval_strategy="epoch",
                            save_strategy="epoch",
                            load_best_model_at_end=True,
                            report_to="none")
```

The `weight_decay` parameter is a regularization technique designed to solve the overfitting problem by penalizing large weights, thereby encouraging them to remain small. This helps control the gradients and can prevent issues like exploding gradients. Weight decay is applied to the model's learnable parameters (i.e., weights that capture data features). However, it is not applied to bias terms or normalization layers. Biases primarily shift the decision boundary, and normalization layers are used to keep the weights centered around zero, making regularization unnecessary for these components.

The `load_best_model_at_end` parameter determines whether the best model, rather than the last one, is returned at the end of training. During training, the model is saved at various checkpoints. The model at the final checkpoint might not always perform best, as it could have worse metrics (such as accuracy or loss) compared to earlier checkpoints. When set to `True`, the best model, based on a specified evaluation metric, is loaded at the end of training. For this feature to work, the `save_strategy` parameter must be configured to specify when checkpoints are saved. Its possible values are:

1. *epoch*: Save at the end of each epoch.

2. *steps* (default): Save every `save_steps`.

3. *best*: Save when a new best metric found.

4. *no*: Create no checkpoints.

The `eval_strategy` determines when the model is evaluated during training. Its possible values are *epoch*, *steps*, and *no* (default) to avoid evaluating the model during training. If set to `steps`, then evaluation is done every `eval_steps`.

The `output_dir` parameter specifies the directory where model checkpoints, training logs, and other outputs (such as evaluation results) are saved during training.

When training a model on multiple devices (e.g., GPUs), it is recommended to use the `per_device_train_batch_size` and `per_device_eval_batch_size` parameters to specify the batch size for each device.

After defining the training hyperparameters, the next step is to define the trainer as an instance of the `transformers.Trainer` class.

```python
import evaluate
metric = evaluate.load("accuracy")

import numpy
def compute_metrics(eval_pred):
    logits, labels = eval_pred
```

```
    predictions = numpy.argmax(logits, axis=-1)
    return metric.compute(predictions=predictions, references=labels)

from transformers import Trainer
trainer = Trainer(model=model,
                  args=training_args,
                  train_dataset=tokenized_datasets["train"],
                  eval_dataset=tokenized_datasets["train"],
                  compute_metrics=compute_metrics)
```

The `model` parameter specifies the model to be used for training, which is typically returned by the `AutoModelForSequenceClassification` class. The training arguments are provided through the `args` parameter.

It's important to note that in this example, the same dataset is assigned to both the `train_dataset` and `eval_dataset` parameters, meaning the same data is used for both training and evaluation. However, this approach is not recommended in practice, as it can lead to overfitting and unrealistic evaluation results.

The `compute_metrics` parameter accepts a function, `compute_metrics()`, which is used to calculate the evaluation metrics during training. This function will receive an object of the `EvalPrediction` class, containing the model's predictions and the corresponding labels. The function must return a dictionary containing the computed evaluation metric(s), such as accuracy in this example.

Once the model is trained, you can save it locally to reuse later for inference. This line saves the trained model to the specified directory (*sms_model* in this case), allowing us to load it later to make predictions.

```
trainer.save_model("sms_model")
```

This log message displays the evaluation loss and accuracy after the final epoch. The model achieved an accuracy of 99.21%. It is important to note that the same dataset is used for both training and evaluation. Figure 4-3 illustrates the evaluation loss and accuracy across all 10 epochs.

```
{'eval_loss': 0.03706911951303482, 'eval_accuracy': 0.9921062073914604,
'eval_runtime': 10.0894, 'eval_samples_per_second': 552.463, 'eval_steps_
per_second': 69.083, 'epoch': 10.0}
```

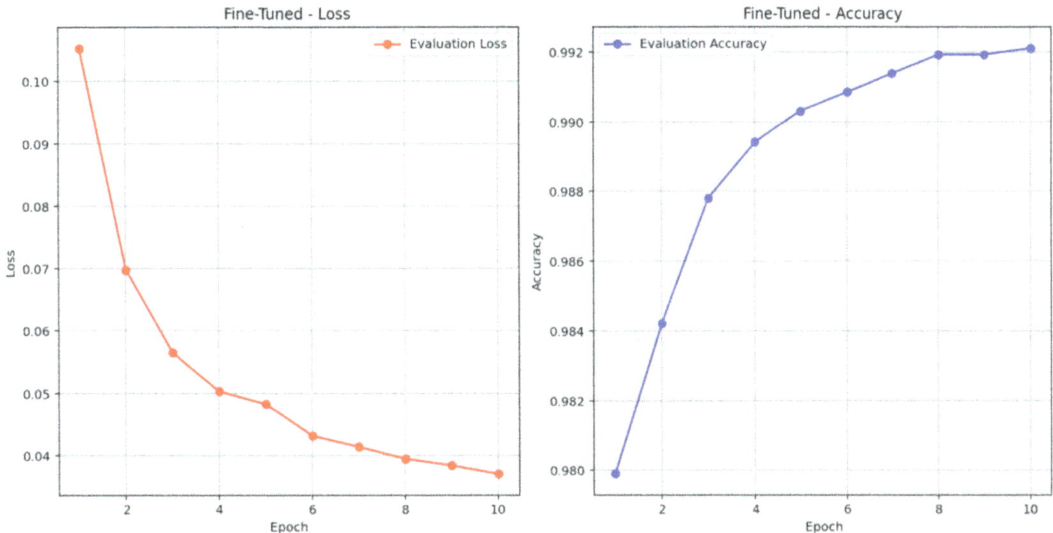

Figure 4-3. *Evaluation loss and accuracy of the fine-tuned model. The classification accuracy increases while the loss decreases over epochs, indicating that fine-tuning is effectively improving the model's performance*

Inference

In the inference phase, we begin by loading the dataset, tokenizer, and trained model from the saved directory, following a process similar to the one used during training.

```python
from datasets import load_dataset
dataset = load_dataset("ucirvine/sms_spam")

from transformers import AutoTokenizer
model_name = 'prajjwal1/bert-tiny'
tokenizer = AutoTokenizer.from_pretrained(pretrained_model_name_or_
path=model_name)

from transformers import AutoModelForSequenceClassification
model = AutoModelForSequenceClassification.from_pretrained("sms_model")
```

The next step is to load a sample from the dataset that will be used for making a prediction.

```python
sample = dataset['train']['sms'][0]
label = dataset['train']['label'][0]
```

The selected sample contains the following text, with a ground-truth label of ham (non-spam).

```
Go until jurong point, crazy.. Available only in bugis n great world la e
buffet... Cine there got amore wat...
```

We then use the tokenizer to convert the loaded sample into token IDs and attention masks, preparing it for input into the model.

```
sample_tokenized = tokenizer.encode(sample, return_tensors="pt")
```

Using the pre-trained model, we obtain the logits, which represent the raw, unnormalized prediction scores for each class.

```
import torch
pred_logits = model(sample_tokenized).logits
```

The logits are then used to determine the predicted class ID by selecting the index with the highest score. Provided as a configuration parameter to AutoModelForSequenceClassification during training, we can convert the predicted class ID into its corresponding label using the id2label mapping.

```
pred_label = torch.argmax(pred_logits).tolist()
pred_class = model.config.id2label[pred_label]
```

This example is a spam SMS message that was correctly identified as spam by the fine-tuned model.

```
Congrats! 1 year special cinema pass for 2 is yours. call 09061209465 now!
C Suprman V, Matrix3, StarWars3, etc all 4 FREE! bx420-ip4-5we. 150pm. Dont
miss out!
```

The following is a new SMS message, not present in the training data. The model classified it as ham (non-spam).

```
You have a very good offer. Contact us soon to claim it. Reach us!
```

When a phone number was added to the SMS, the model classified it as spam. This suggests that the model has learned that spam messages frequently include a contact method, such as a phone number, to target the recipient.

```
You have a very good offer. Contact us soon to claim it. Reach 111111111!
```

This concludes our first fine-tuning example, which demonstrates full fine-tuning to adapt a base pre-trained model to a specific target task.

4.2.2 Parameter-Efficient Fine-Tuning (PEFT)

Full fine-tuning provides an efficient alternative to training a model from scratch, which is computationally expensive. This approach uses a pre-trained base model and updates all of its parameters to adapt it to a new task or dataset. Ideally, fine-tuning improves the model's performance in the target domain while preserving and building upon the broad knowledge already encoded in the base model.

Full fine-tuning adjusts all of the model's parameters, progressively shifting the model away from its original training distribution. This presents a double-edged sword. If the new dataset is large and of high quality, this shift can be beneficial to maximize the model's ability to specialize in the target task. But if the new data is limited or noisy, it may degrade performance by causing the model to forget previously learned knowledge (i.e., *catastrophic forgetting*).

Beyond the risk of forgetting, fine-tuning also poses significant computational challenges. Large models contain billions of parameters, and updating all of them, like the weight matrices in attention and feedforward layers, is computationally expensive and time-consuming.

For instance, in a model like GPT-3 (davinci), which has an embedding size of $d_{\text{model}} = 12,288$, 96 attention heads, a head dimension of 128 for each weight matrix, and 96 attention layers, its total number of trainable parameters exceeds 175 billion. As a result, full fine-tuning of such a large number of parameters is impractical for many organizations with limited infrastructure, motivating the exploration of more efficient adaptation methods through parameter-efficient fine-tuning (PEFT).

PEFT is a practical way to fine-tune large models while significantly reducing the number of trainable parameters. Instead of updating the entire parameter space, PEFT selectively fine-tunes a small subset. For instance, it may focus only on the weights associated with the query matrices in the attention mechanism. In some large language models, the number of parameters tied to the query matrices across all attention layers and heads can be as high as

$$12,288 \times 128 \times 96 \times 96 \approx 14\text{B}, \quad (1)$$

which accounts for approximately 8% of the total model parameters. For the BERT-base, which contains roughly 110 million trainable parameters. With a model dimension $d_{model} = 768$, 12 attention layers, 12 attention heads, and a head dimension of 64, the number of parameters associated with the query matrices is

$$768 \times 64 \times 12 \times 12 \approx 7M, \qquad (2)$$

representing around 6% of the model's total parameters. This smaller parameter subset can be trained efficiently on modest hardware, making fine-tuning accessible to a wider range of researchers and practitioners.

PEFT models are both compact and extensible. *Compactness* refers to the fact that only a limited number of parameters are fine-tuned, reducing computational and storage requirements. *Extensibility* means that the base model's learned knowledge is not overwritten, but rather extended, since most of its original parameters remain unchanged during fine-tuning.

If you have sufficient computational resources and a large high-quality dataset, full fine-tuning is the best option, as it allows the model to adjust all its parameters to better fit the new data. However, when data or resources are limited, partial fine-tuning provides a viable alternative. While it may not achieve the same level of performance as full fine-tuning, it serves as an effective starting point for adapting large models with reduced computational cost.

Imagine you love blackberry cake, but it's expensive to buy, and you do not have enough time to prepare it from scratch. However, the nearest bakery offers a cheap strawberry cake. So, you managed to buy the strawberry cake along with a small amount of blackberries to modify the cake and infuse it with some of your preferred blackberry flavor.

Now, think of the base model as the strawberry cake, as shown in Figure 4-4A. You want to adapt this model to your custom data, represented by the blackberries.

If you apply transfer learning, a step that usually accompanies fine-tuning, by modifying only the top layers of the model, it's like placing a few blackberries on top of the strawberry cake (Figure 4-4B). The core remains the same, but with a slight touch of your desired flavor.

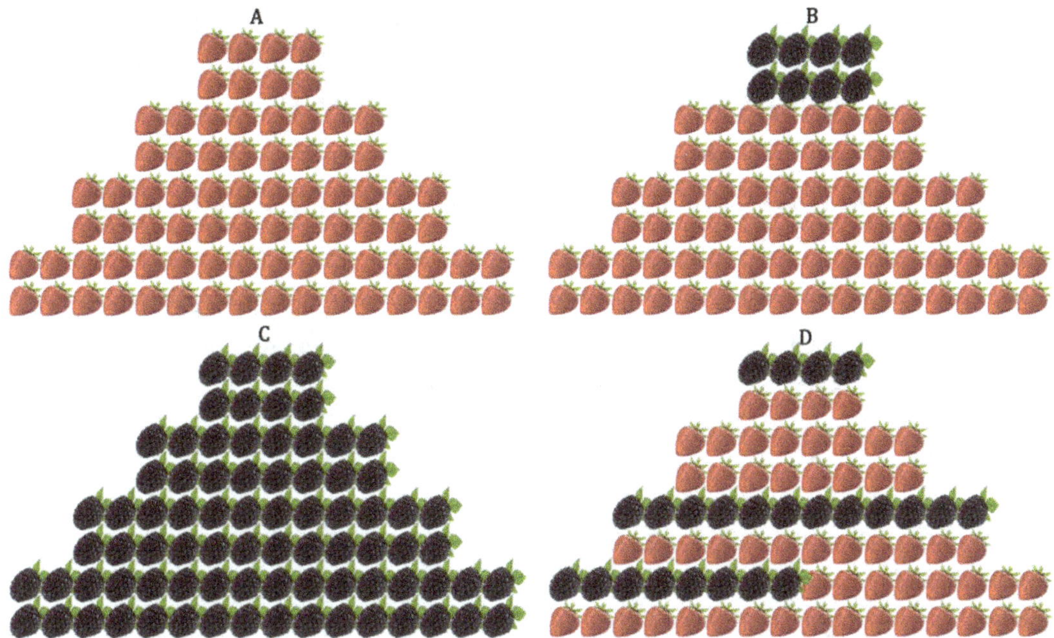

Figure 4-4. *Variants of model fine-tuning. (A) The base model without any modifications. (B) Transfer learning by replacing and training only the top layers. (C) Full fine-tuning where all model parameters are updated during training. (D) Partial fine-tuning involving updates to only specific components of the model*

In Figure 4-4C, full fine-tuning is analogous to replacing every strawberry in the cake with blackberries, completely transforming its flavor. However, this requires many blackberries, just as full fine-tuning demands a large amount of data to effectively change the model's behavior. With only a few blackberries, such a transformation wouldn't be feasible.

Finally, Figure 4-4D illustrates partial fine-tuning, where only some of the strawberries are swapped for blackberries. The resulting cake still tastes mostly like the original, but with a gentle shift toward your preference in some restricted parts of the cake. While the strawberry flavor still dominates, the outcome is acceptable, especially considering you didn't spend much money on the more expensive blackberry cake.

Two popular PEFT techniques are adapters and LoRA (low-rank adaptation). While they use different strategies, both aim to reduce the number of trainable parameters during partial fine-tuning while maintaining performance comparable to full fine-tuning.

> *Adapters and LoRA represent distinct approaches within the broader family of PEFT methods.*

QLoRA (Quantized LoRA) is an optimized variant of LoRA designed to further reduce memory usage by applying quantization, making it especially suitable for fine-tuning large models on resource-constrained hardware.

Adapters

The adapter's adaptation method works by plugging an external FFNN, called adapter, into the model while freezing all the model parameters. The parameters of the new adapters are the only trainable parameters. According to Figure 4-5, there are two adapters added for each attention block:

1.   After the multi-head attention layer

2.   After the MLP (i.e., feedforward) layer

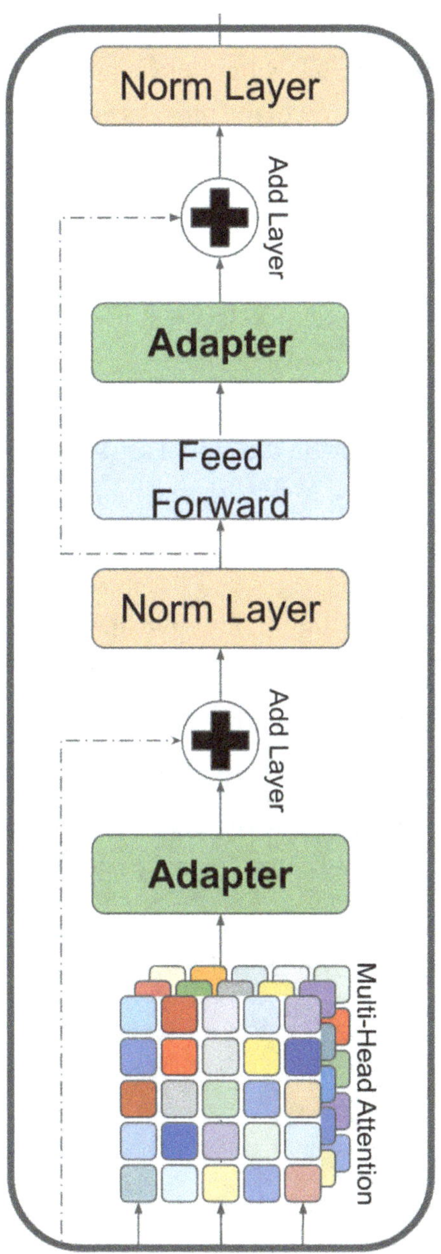

Figure 4-5. *Integration of PEFT adapter modules into the transformer architecture. Two adapters are inserted into each transformer block. The first one is after the multi-head attention layer, and the second one is after the feedforward neural network*

The modified Transformer model, with the adapter module inserted, should initially behave similarly to the original model without the adapter. Suppose that the input to the adapter is derived from the output of the model's existing parameters W and input X, the adapter's parameters $W_{adapter}$ are initialized to approximate an identity function. This ensures that, at the beginning of training, the adapter does not significantly alter the model's behavior.

$$f_{W_{adapter}}\left(f_W\left(X\right)\right) \approx f_W\left(X\right) \quad (3)$$

During fine-tuning, only the adapter's parameters are updated based on the new data, while all other parameters in the model are kept frozen. This preserves the pre-trained knowledge encoded in the original model and prevents it from being overwritten. As a result, the learning from new data is effectively restricted to the newly introduced adapter layers.

The adapter FFNN receives an input vector of length d_{model}, which goes through the architecture. The adapter consists of three layers, as illustrated in Figure 4-6.

1. **Linear (Down-Projection)**: Reduces the input dimensionality from the original size d_{model} to a smaller intermediate dimension $d_{adapter}$.

2. **Non-Linear Activation**: Applies a non-linear activation function (such as ReLU) to introduce non-linearity and capture complex patterns in the data.

3. **Linear (Up-Projection)**: Projects the transformed vector back to the original dimensionality d_{model}.

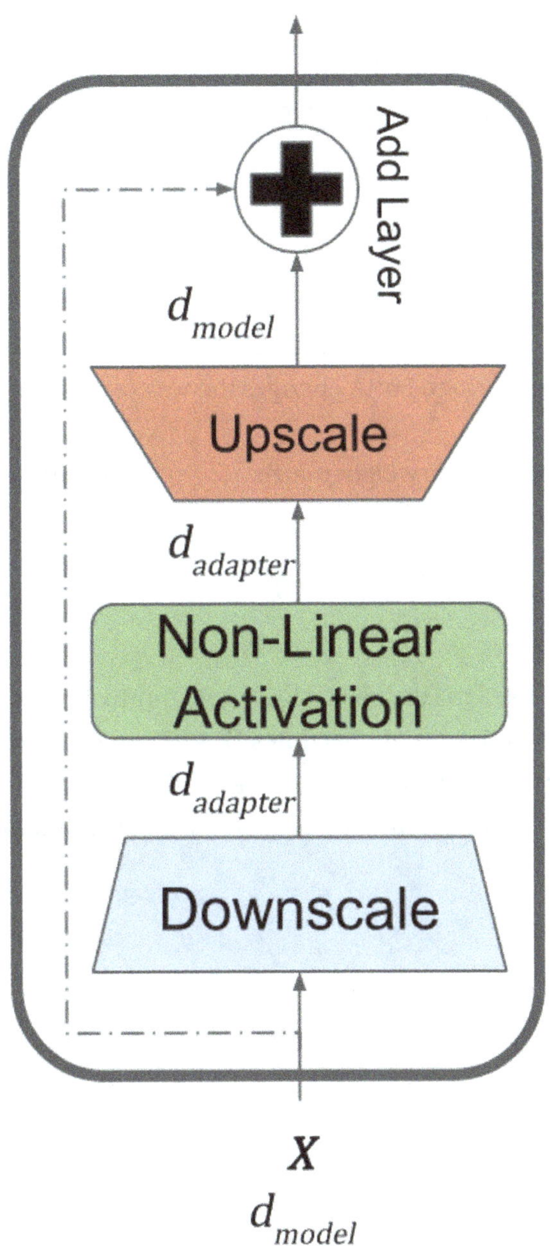

Figure 4-6. *Architecture of the adapter feedforward neural network. The input embedding vector undergoes a downscaling operation followed by an upscaling operation, reducing the total number of trainable parameters while enabling efficient adaptation*

This down-projection and up-projection structure significantly reduces the number of trainable parameters during fine-tuning, allowing the model to adapt to the new data by training few parameters. Remember that the base model's parameters are already frozen.

LoRA

LoRA (low-rank adaptation) is a PEFT technique that differs from the adapter by avoiding the addition of new layers to the model architecture. Instead of modifying the structure of the Transformer, LoRA changes the way certain weight matrices are updated.

Rather than updating a large weight matrix $W \in \mathbb{R}^{d_{\text{model}} \times d_{\text{model}}}$ directly, LoRA introduces two smaller trainable matrices:

- $A \in \mathbb{R}^{r \times d_{\text{model}}}$

- $B \in \mathbb{R}^{d_{\text{model}} \times r}$

These matrices are used to construct a low-rank approximation of the weight update, such that the effective weight becomes $W + BA$, as illustrated in Figure 4-7. The hyperparameter r represents the rank or bottleneck dimension, controlling the trade-off between efficiency and reduction in the number of trainable parameters.

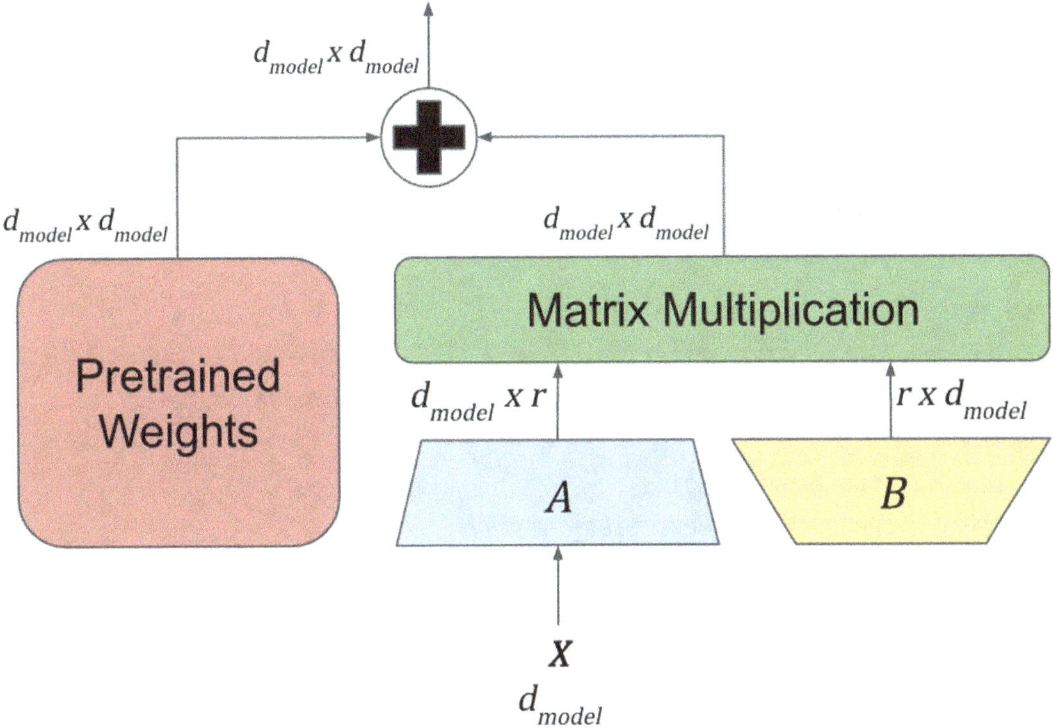

Figure 4-7. *LoRA PEFT technique*

For example, consider a weight matrix $W \in R^{512 \times 512}$, which contains a total of 262,144 trainable parameters. Using LoRA, we replace this full matrix update with two much smaller trainable matrices:

- $A \in R^{512 \times 2}$

- $B \in R^{2 \times 512}$

Together, these matrices introduce only $512 \times 2 + 2 \times 512 = 2,048$ trainable parameters, which is just 0.78% of the original parameter count in W. By multiplying A and B, we construct a low-rank matrix $\tilde{W} = AB$ of the same shape as W, and this is added to the frozen pre-trained weights:

$$W_{new} = W_{old} + AB \qquad (4)$$

These low-rank matrices capture the knowledge from the new data, while the original weights W_{old} remain untouched. This approach shifts the behavior of the base model toward the new task, without overwriting the pre-trained knowledge.

Here is an example of using LoRA in the Hugging Face peft library. Make sure it is already installed.

As we did before for fine-tuning all the parameters, the dataset, tokenizer, and model (prajjwal1/bert-tiny) are loaded.

```python
from datasets import load_dataset
dataset = load_dataset("ucirvine/sms_spam")

from transformers import AutoTokenizer
model_name = 'prajjwal1/bert-tiny'
tokenizer = AutoTokenizer.from_pretrained(pretrained_model_name_or_
path=model_name)
def tokenize_function(examples):
    return tokenizer(text=examples["sms"],
                     padding="max_length",
                     max_length=186,
                     truncation=True)
tokenized_datasets = dataset.map(tokenize_function, batched=True)

id2label = {0: "ham",
            1: "spam"}

label2id = {v: k for k, v in id2label.items()}


from transformers import AutoModelForSequenceClassification
model = AutoModelForSequenceClassification.from_pretrained(pretrained_
model_name_or_path=model_name,
                                        num_labels=2,
                                        id2label=id2label,
                                        label2id=label2id)
```

Rather than passing the model directly to the `Trainer` object, it is first adapted using LoRA to modify the trainable parameters according to the newly introduced LoRA components. This process involves two main steps:

1. Create an instance of the `peft.LoraConfig` class to specify the configuration parameters for LoRA.

2. Use the `peft.get_peft_model()` function to generate the LoRA-adapted version of the model based on the previously defined `LoraConfig` object.

This is how the `LoraConfig` object is created.

```python
from peft import LoraConfig

loca_config = LoraConfig(task_type='SEQ_CLS',
                r=4,
                lora_alpha=32,
                lora_dropout=0.1,
                target_modules=['query'])
```

The `lora_alpha` parameter serves as a scaling factor (can be considered a learning rate) applied to the learned LoRA weights before they are added to the original model weights. Without this parameter, the updated weight matrix W' is computed as:

$$W' = W + \Delta W = W + BA \qquad (5)$$

When `lora_alpha` is specified, it scales the product of the low-rank matrices B and A, effectively controls the influence of the LoRA adaptation on the base model:

$$W' = W + \Delta W = W + \alpha(BA) \qquad (6)$$

A larger value of `lora_alpha` increases the contribution of the LoRA weights, allowing the model to adapt more rapidly to the new training data. However, this also increases the risk of the model deviating too quickly from its original pre-trained knowledge. Conversely, a smaller value diminishes the impact of the LoRA update, making less adaptation.

The `r` parameter refers to the rank of the two matrices used to form the original frozen weight matrix W of size $x \times y$. If $r = 4$, then A and B will have rank-4 matrices of sizes $x \times 4$ and $4 \times y$, respectively. To determine an appropriate rank value, consider the relevance of the new data to the data used to train the base model. If the new data closely aligns with the original training data, a lower rank may be sufficient. However, if the new data introduces different information, a higher rank is recommended to effectively incorporate the new knowledge into the model.

The `lora_dropout` parameter sets the dropout rate applied to the trainable low-rank matrices A and B during training. This regularization technique helps prevent overfitting on the new dataset by randomly zeroing some of the weights, thereby creating more robust learning.

Additionally, the `task_type` parameter indicates the nature of the task for which LoRA is being applied. This helps the fine-tuning library (such as Hugging Face's `peft`) correctly configure the model. Common task types include:

- `CAUSAL_LM`: Causal Language Modeling

- `SEQ_CLS`: Sequence Classification

- `TOKEN_CLS`: Token Classification

- `QUESTION_ANS`: Question Answering

- `SEQ_2_SEQ_LM`: Sequence-to-Sequence tasks (e.g., translation, summarization)

- `FEATURE_EXTRACTION`: Feature Extraction

For a full, up-to-date list of task types, refer to the PEFT source code `https://github.com/huggingface/peft/blob/2f063e6342235842a00bca150a98854b89117986/src/peft/utils/peft_types.py#L69`.

To determine the appropriate values for the `target_modules` parameter, simply iterate over the named modules within the target model. These module names vary between models, so it is important to inspect them directly. If the model is implemented in PyTorch and you are specifically interested in adapting only the linear layers, you can filter the modules by checking whether the word `"Linear"` appears in the module's type name. This is because linear layers in PyTorch are instances of the `torch.nn.modules.linear.Linear` class.

```python
from transformers import AutoModel

model = AutoModel.from_pretrained("distilbert-base-uncased")

for name, module in model.named_modules():
    if "Linear" in str(type(module)):
        print(name, "->", type(module))
```

The naming of attention-related modules varies across different transformer architectures. For DistilBERT (`distilbert-base-uncased`), the query, key, and value projection layers are named `q_lin`, `k_lin`, and `v_lin`, respectively. The output projection layer is referred to as `out_lin`. This is the same for GPT-2.

In contrast, the base BERT model (`bert-base-uncased`) follows a more descriptive naming convention: `query`, `key`, `value`, and `output.dense`. This same convention is used in variants such as `prajjwal1/bert-tiny`.

For the T5 architecture (e.g., `t5-small` from Google), the corresponding module names are further abbreviated: `q`, `k`, `v`, and `o`, denoting query, key, value, and output projections, respectively.

Note that the term *module* is used instead of *layer* because it provides a broader and more accurate description of a model's components. In PyTorch, layers such as `Linear` are subclasses of the more general `Module` class, making *module* the preferred term when referring to parts of the model to be fine-tuned.

Once the `LoraConfig` object is defined, the next step is to create the LoRA version of the model using the `peft.get_peft_model()` function.

```
from peft import get_peft_model
peft_model = get_peft_model(model, lora_config)
```

The command `peft_model.print_trainable_parameters()` displays the number of trainable parameters in the LoRA-adapted model relative to the total number of parameters in the original base model. In this case, only 2,306 parameters are marked as trainable, representing only 0.0525% of the full model parameters, highlighting how LoRA significantly reduced the number of parameters to fine-tune.

```
trainable params: 2,306 || all params: 4,388,484 || trainable%: 0.0525
```

Once the PEFT LoRA model is created, the next step is to initialize a `Trainer` object to train the model as usual. After training, the resulting model is saved for later use. The number of training epochs is now 50, instead of just 10 as in the previous full parameter fine-tuning example.

```
from transformers import TrainingArguments, Trainer
training_args = TrainingArguments(output_dir="sms_fine_tuning_LoRA",
                                  learning_rate=0.00001,
                                  num_train_epochs=50,
                                  weight_decay=0.01,
```

```python
                                        eval_strategy="epoch",
                                        save_strategy="epoch",
                                        load_best_model_at_end=True,
                                        report_to="none")

import numpy
import evaluate
metric = evaluate.load("accuracy")

def compute_metrics(eval_pred):
    logits, labels = eval_pred
    predictions = numpy.argmax(logits, axis=-1)
    return metric.compute(predictions=predictions, references=labels)

trainer = Trainer(model=peft_model,
                  args=training_args,
                  train_dataset=tokenized_datasets['train'],
                  eval_dataset=tokenized_datasets['train'],
                  compute_metrics=compute_metrics)

trainer.train()

trainer.save_model("sms_model_loRA")
```

The following log message reports the evaluation loss and accuracy after the final epoch. Figure 4-8 presents the evaluation loss and accuracy over the 50 training epochs of the LoRA-adapted model.

```
{'eval_loss': 0.06438042223453522, 'eval_accuracy': 0.9817007534983854,
'eval_runtime': 10.7541, 'eval_samples_per_second': 518.316, 'eval_steps_
per_second': 64.813, 'epoch': 50.0}
```

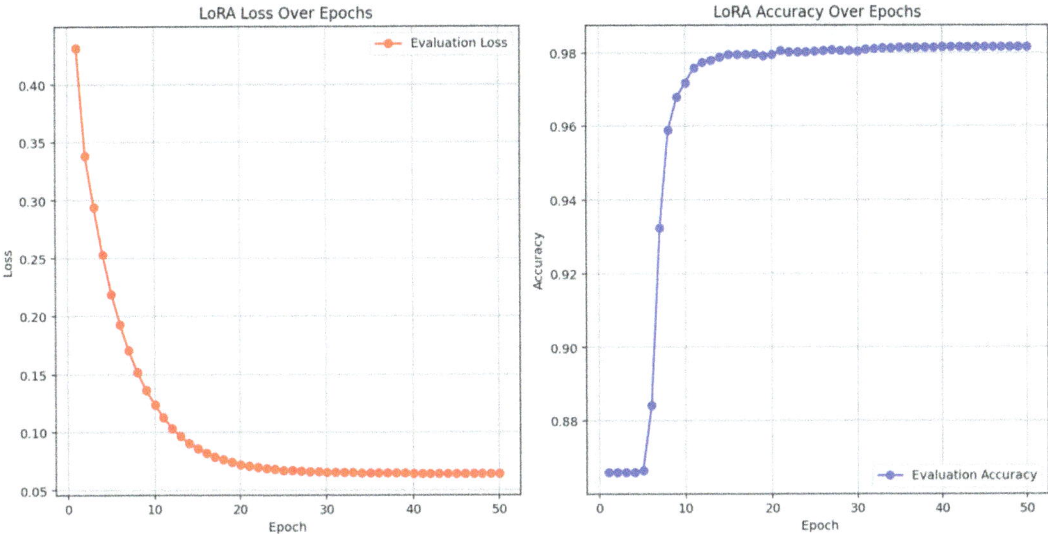

Figure 4-8. *Evaluation loss and accuracy of the LoRA fine-tuned model. The decreasing loss and increasing accuracy over epochs indicate that the model is learning effectively while maintaining efficiency by training only a small subset of parameters*

The model achieved an accuracy of 98.17%, compared to 99.21% obtained when all parameters were fine-tuned. This result was achieved using 50 training epochs with the LoRA model, as opposed to just 10 epochs for the full model. This highlights a key trade-off introduced by LoRA.

While the LoRA PEFT method significantly reduces the number of trainable parameters, enabling the fine-tuning of large models with millions or even billions of parameters on limited resources, it does so at the cost of reduced model performance. Moreover, with fewer parameters, the model requires more training epochs to approximate the performance of the fully fine-tuned model.

As a conclusion, instead of incurring the additional cost of fine-tuning a massive model with billions of trainable parameters, just reduce the cost by creating an approximation model with only thousands or a few million trainable parameters.

4.2.3 Text Generation

Fine-tuning a model for text generation generally follows a pipeline similar to the one used with sequence classification, with two primary differences:

1. The training data

2. The model being fine-tuned

Even without fine-tuning, a generative model can produce coherent text. However, to observe a noticeable change in the generated text style, the fine-tuning dataset must introduce a distinctive feature. This feature could be a specific tone, writing style, or domain-specific knowledge that the original model was not trained on. By incorporating such data, the fine-tuned model can produce outputs that differ from those of the base model.

We will use synthetic data generated by prompting a pre-trained model to act as a cat. The model is instructed to produce a set of prompt-response pairs that use some cat-like expressions, such as *meow* and *hiss*. This data will be used to fine-tune the model, resulting in outputs that reflect the intended behavior. Here are some samples from the data.

```
{"prompt": "What is your primary goal in life?",
 "response": "Meow! To nap in all the sunbeams. Hiss! And to ensure my
human provides an endless supply of tuna."}

{"prompt": "Why do you meow at closed doors?",
 "response": "Meow! Doors are a challenge, a barrier to be overcome. Hiss!
Or perhaps there's a hidden tuna room on the other side."}
```

The data is saved in the `cat_data.json` file, which can be loaded using the `load_dataset()` function in the Hugging Face `datasets` library.

```
from datasets import load_dataset
dataset = load_dataset("json", data_files="cat_data.json")
```

This is the returned dataset dictionary. We are not going to use the prompt for now.

```
DatasetDict({
    train: Dataset({
        features: ['prompt', 'response'],
```

```
        num_rows: 583
    })
})
```

The model that will be fine-tuned is `EleutherAI/gpt-neo-125m`. It has just 125 million parameters, which is somehow manageable to be fine-tuned. Using the `AutoTokenizer` class, the tokenizer is automatically loaded.

```
from transformers import AutoTokenizer
model_name = 'EleutherAI/gpt-neo-125m'
tokenizer = AutoTokenizer.from_pretrained(pretrained_model_name_or_
path=model_name)
```

The returned tokenizer is an instance of the `GPT2TokenizerFast` class. As the name suggests, it is an optimized, faster version of the original GPT-2 tokenizer. By default, this tokenizer does not include a padding token, since GPT-2 models were not originally trained with one.

The `EleutherAI/gpt-neo-125m` model is a decoder-only model designed to generate text one token at a time, typically from a single input sequence. By design, it does not require padding during inference, as it processes one sequence at a time. However, during training, sequences within a batch must have equal lengths. To enable padding functionality, a padding token `'<|endoftext|>'` must be explicitly added by setting the tokenizer's `pad_token` property.

```
tokenizer.pad_token = tokenizer.eos_token
```

You can find the padding token ID by either looking into the dictionary:

```
tokenizer.convert_tokens_to_ids(tokenizer.pad_token)
```

Or by directly using the `pad_token_id` property:

```
tokenizer.pad_token_id
```

It returns 50256. In the encoded text sequence, the number 50256 is used to indicate a padding token. The `tokenize_function()` helper function handles the padding, where the tokenizer is configured with `padding="max_length"`. Although the model's default maximum sequence length is 1024 tokens, this value is reduced to 128 in our case to accommodate the short sequences present in the dataset.

```python
def tokenize_function(examples):
    return tokenizer(text=examples["response"],
                     padding="max_length",
                     max_length=128,
                     truncation=True)
tokenized_datasets = dataset.map(tokenize_function, batched=True)
```

We are going to use the accuracy as the evaluation metric. This works by comparing each predicted token at each step against the actual text. This ensures that the same exact token is predicted at all times. This is not the best metric, but it is good as a starting point. For the accuracy to be measured, it must use a label. This is why the add_labels() function is created to add a new column called labels, which has the same token IDs generated for the response column.

By default, padding tokens with the ID 50256 are treated as regular tokens during loss computation, which can negatively impact the model's performance by forcing it to predict padding tokens. To address this, the add_labels() function not only adds labels to the dataset but also replaces padding tokens with the special token ID -100, ensuring that they are automatically ignored by the Hugging Face trainer.

```python
def add_labels(example):
    example["labels"] = [-100 if token_id == tokenizer.pad_token_id else
    token_id for token_id in example["input_ids"]]
    return example
tokenized_datasets = tokenized_datasets.map(add_labels)
```

As usual, the model is loaded and the training arguments are prepared. The AutoModelForCausalLM class is used to automatically load the right class for the model by its name. In this case, the returned model is an object of the GPTNeoForCausalLM class.

```python
from transformers import AutoModelForCausalLM
model = AutoModelForCausalLM.from_pretrained(pretrained_model_name_or_
path=model_name)
```

```python
from transformers import TrainingArguments, Trainer
training_args = TrainingArguments(output_dir="cat_fine_tuning",
                                  learning_rate=0.00001,
                                  num_train_epochs=10,
                                  weight_decay=0.01,
```

```
        eval_strategy="epoch",
        save_strategy="epoch",
        load_best_model_at_end=True,
        remove_unused_columns=True,
        report_to="none")
```

The compute_metrics() function measures the accuracy between the predicted and actual responses, which is used by the trainer for evaluating the model. The predictions and labels arrays are flattened just to remove the unnecessary dimension representing the sample ID. It is useless for accuracy. Since we have 583 samples, the size of these arrays is (583, 128). Since the dimension is not needed, the flattened array size is (74624,).

```python
import numpy
import evaluate
metric = evaluate.load("accuracy")

def compute_metrics(eval_pred):
    logits, labels = eval_pred
    predictions = numpy.argmax(logits, axis=-1)

    predictions = predictions.flatten()
    labels = labels.flatten()

    mask = labels != -100
    predictions = predictions[mask]
    labels = labels[mask]

    predictions = predictions.astype(numpy.int32)
    labels = labels.astype(numpy.int32)

    return metric.compute(predictions=predictions, references=labels)
```

The model must measure the prediction accuracy over each token in the sequence, including the padding tokens that degrade its performance. To ignore the padding tokens while calculating the model's accuracy, we have to ignore them by explicitly filtering out the padding tokens with ID -100. This reduces the flattened array size to just (11432,).

Once the model, training arguments, data, and evaluation metrics are defined, the trainer object is created. The trained model is saved in the `cat_model` directory.

```
trainer = Trainer(model=model,
                  args=training_args,
                  train_dataset=tokenized_datasets['train'],
                  eval_dataset=tokenized_datasets['train'],
                  compute_metrics=compute_metrics)

trainer.train()

trainer.save_model("cat_model")
```

Inference

The trained model is loaded through a pipeline. Note that the same tokenizer of the `EleutherAI/gpt-neo-125m` model is used because we did not change the tokenizer.

```
from transformers import AutoTokenizer
model_name = 'EleutherAI/gpt-neo-125m'
tokenizer = AutoTokenizer.from_pretrained(pretrained_model_name_or_
path=model_name)

from transformers import AutoModelForCausalLM
model = AutoModelForCausalLM.from_pretrained("cat_model")

from transformers import pipeline
catmodel = pipeline("text-generation",
                    model="cat_model",
                    tokenizer=tokenizer)
```

This is a sample to ask the model to generate something crazy cats do.

```
sample = "Give one crazy thing cats do"
prediction = catmodel(sample)
```

This is the model's answer. It was able to capture the salient properties of the training data by using some common words like *meow* and *hiss*.

```
Give one crazy thing cats do! 'Meow'-sterious chase! Hiss... what's that?
```

A fundamental limitation of standard language models is that they generate text by sequentially predicting the next word, rather than directly addressing a given question or prompt. This approach can lead to responses that merely extend the input rather than providing a meaningful answer. Since the model progressively generates text, it may repeat parts of the prompt even when they do not logically fit within the response.

For instance, if prompted with *"Give one crazy thing cats do,"* a model that lacks task adaptation may not recognize this as a question requiring an answer. Instead, it might begin its response by repeating the prompt. An appropriate response should start with something like, *"One of the unusual behaviors of cats is..."* However, without explicit instruction, the model may fail to recognize that the intent behind the prompt is to answer a question.

This issue arises because the model has not been explicitly trained for question-answering tasks. In self-supervised learning, the model is trained to generate the expected output based solely on the provided input. However, to enhance its ability to follow specific instructions, instruction fine-tuning is introduced.

Instruction fine-tuning adapts the model for task-specific objectives by using explicit instructions alongside input-output pairs. Instead of simply providing the prompt *"Give one crazy thing cats do,"* the model is given a directive such as, *"Answer the following question."* This additional instruction guides the model to understand the task more effectively and produce more relevant responses.

4.3 Prompt Engineering

You've likely used a generative AI tool such as OpenAI's ChatGPT or Anthropic's Claude by providing it with a prompt. Often, the first response may not fully meet your expectations. In such cases, you might revise the prompt to correct misunderstandings, fill in missing details, or enhance the overall quality of the output. This iterative process is known as prompt engineering.

> *Engineering is generally about building things that work effectively.*
>
> *Prompt engineering is about crafting prompts that lead AI models to produce the desired results.*

At its core, prompt engineering is the practice of designing prompts that effectively guide a generative AI model toward producing useful, accurate, or creative outputs.

Suppose you want the AI to generate a Python script that finds a route between two locations. You might start with the following prompt:

```
Write a Python script that receives two locations and returns the route.
```

After reviewing the output, you notice that the route uses highways by default. If your goal is to avoid highways, you would revise the prompt to make that requirement explicit:

```
Write a Python script that receives two locations and returns the route
while avoiding highways.
```

Through such revisions, you iteratively refine the prompt until the model produces the desired output.

Sometimes, the model cannot determine the best response based on a general instruction alone. In these cases, it's helpful to provide additional guidance in the prompt, such as examples of input-output pairs. This technique is known as in-context learning, where the model learns the task by observing examples provided directly in the prompt.

In-context learning is especially useful when the task is complex or ambiguous or requires a specific style or structure. Rather than relying solely on instructions, you show the model *how* to respond by demonstration.

4.3.1 In-Context Learning (ICL)

In some cases, we may have limited or no training data available to fine-tune a model. This makes traditional fine-tuning techniques impractical. In-context learning (ICL) addresses this challenge by enabling models to adapt to a target task, as in Figure 4-9. It uses either no examples (*zero-shot learning*), one sample (*one-shot learning*), or a small number of examples (*few-shot learning*). This is a kind of prompt engineering that aims to create a better prompt for solving a task with no or few examples.

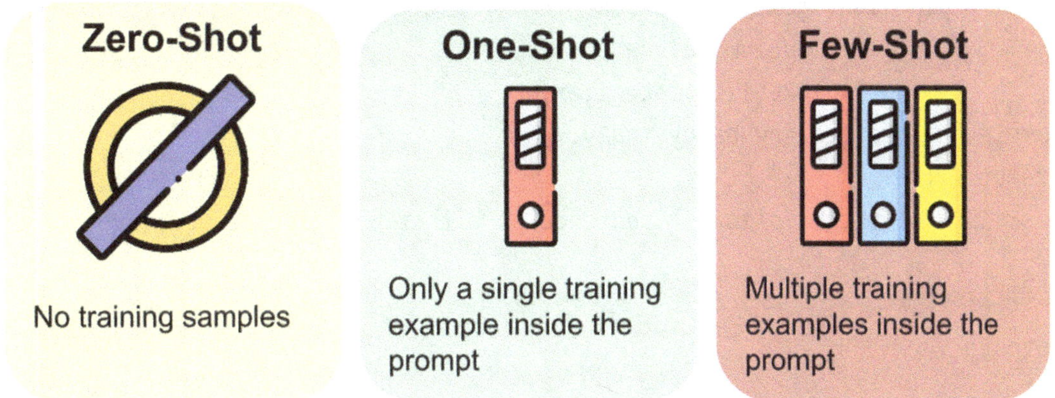

Figure 4-9. Paradigms of in-context learning are zero-shot, one-shot, and few-shot

Zero-Shot

In zero-shot learning, the model is guided through detailed instructions embedded within the prompt. These instructions provide context about the task and clarify the expected behavior of the model. For example, instead of simply providing the model with the following prompt:

```
I am feeling happy today.
```

We can include an explicit instruction within the prompt to clarify the task, such as:

```
Classify the following sentence as either positive or negative
```

If the base model is trained enough to adapt quickly to specific tasks, it will understand the purpose is to return either positive or negative given the sentiment of the input. This is an example that uses the `EleutherAI/gpt-neo-1.3B` model.

```python
from transformers import AutoTokenizer
model_name = 'EleutherAI/gpt-neo-1.3B'
tokenizer = AutoTokenizer.from_pretrained(pretrained_model_name_or_
path=model_name)
```

```python
from transformers import pipeline
model = pipeline("text-generation",
                 model=model_name,
                 tokenizer=tokenizer)
sample = "I am feeling happy today."
prompt = f"""
Task: Classify the following sentence as either positive or negative.

{sample}

Output:
"""

output = model(prompt, max_new_tokens=10)
print(output[0]["generated_text"])
```

Without any examples, the model's responses typically failed to address the question and often returned irrelevant answers. On one occasion, it simply repeated the input sentence. On another occasion, it added the word *Positive* in front of the input sentence, as shown in the following response. Although the model occasionally produces outputs that are closer to the expected result, its performance remains inconsistent and inaccurate.

```
Positive: I am feeling happy today.
Negative
```

One-Shot

In one-shot learning, only a single sample is provided to the model through the prompt. This sample includes both the input and the corresponding ground-truth response, guiding the model on how to generate future responses. The new prompt contains just one example, which explicitly instructs the model to return the word *Positive* when the input reflects a positive sentiment. However, it does not include any examples or instructions for handling negative samples.

```python
sample = "I am feeling happy today."
prompt = f"""
Task: Classify the following sentence as either positive or negative.
```

```
Example:
Input: I am feeling great today.
Output: Positive

{sample}

Output:
"""
```

When the model is given the positive example, *"I am feeling happy today,"* it correctly returns the word Positive. Similarly, for the negative example, *"I am feeling bad today,"* the model correctly responded with Negative. This demonstrates that the model learned the task from a single example.

Few-Shot

However, in some cases, a single sample is not enough to fully adapt the model to a new task. In our example, the model still exhibited some drawbacks, as it often generated more text than just the words *Positive* or *Negative*. To address this, few-shot learning can be used, where multiple examples are provided to the model, as in the next example. This approach offers the model more context, helping it understand the expected behavior and produce more accurate responses.

```
sample = "What a nice day!"
prompt = f"""
Task: Classify the following sentence as either positive or negative.

Example 1:
Input: "I am feeling great today."
Output: "Positive"

Example 2:
Input: "I am feeling sad and down."
Output: "Negative"

{sample}

Output:
"""
```

Using ICL, the model can be customized to perform a specific task without the need for fine-tuning. However, there are still some drawbacks. For instance, we must include examples in every prompt to provide the necessary context for the model to adapt to the task. Moreover, ICL is effective only when the examples are clear, and the model has the foundational knowledge to understand them. As the complexity of the examples increases, the model's performance tends to degrade.

One limitation of ICL or prompt engineering in general is the increasing length of prompts, which results in a larger number of input tokens being processed by the model. This can significantly raise costs. In scenarios where numerous long prompts are expected, fine-tuning the model may offer a more cost-effective solution. By embedding the necessary knowledge directly into the model, fine-tuning reduces the need to include extensive contextual information in each prompt.

4.3.2 Chain of Thought Prompting

Chain of thought (CoT) is not only useful for helping humans understand and interpret a model's reasoning process; it also serves as a mechanism for guiding the model to break down complex problems into simpler, more manageable subproblems. CoT represents the model's reasoning steps before producing its final answer to a prompt.

Some models inherently think more deeply by breaking problems into subproblems. Others require explicit prompting to produce a CoT. One effective approach is to provide the model with a few worked examples (i.e., few-shot learning), which it can use as context for generating its own reasoning chains. This is especially helpful when the model will be prompted with similar problems in the future.

For example, consider the question:

What is the total sum of elements in a Python list with three elements `[6, -2, 7]`*?*

A CoT-enabled reasoning process would proceed as follows:

1. Start with the first element (6) in the list.

2. Add the first element (6) to the second element (-2) to get 4.

3. Add this result (4) to the third element (7) to get 11.

4. Return the final sum, 11.

By explicitly guiding the model through these steps, CoT helps ensure correctness and transparency in the reasoning process.

4.4 Quantization

Instead of storing every model weight in its full-length floating-point format (such as 32-bit floating point, FP32), quantization reduces the precision of these weights to lower-bit representations, such as 16-bit floating point (FP16), 8-bit integers (INT8), or even 4-bit integers (INT4). By reducing precision, quantization significantly decreases the model's storage size and memory usage and often accelerates inference, making it possible to run large models on less powerful hardware.

However, this efficiency gain can come at the cost of a slight degradation in model performance, depending on the quantization method, bit width, and sensitivity of certain layers to reduced precision. In practice, the trade-off between performance and efficiency is often acceptable, especially when deploying models in resource-constrained environments.

The Hugging Face `transformers` library supports several quantization techniques, allowing you to specify the precision directly when loading a model. To load a quantized model, all you need to do is pass another parameter to indicate the quantization bits.

```
.from_pretrained(..., load_in_4bit=True)
```

This command loads the model in 4-bit precision by setting `load_in_4bit` to True, enabling models with millions (or even billions) of parameters to fit into the memory of consumer-grade GPUs or even some CPUs. Similarly, `load_in_8bit=True` can be used for 8-bit quantization.

Beyond these options, more advanced methods such as QLoRA (Quantized LoRA) combine low-bit quantization with parameter-efficient fine-tuning, achieving memory efficiency while retaining performance close to that of full-precision models. Quantization-aware training (QAT) is another approach in which the model is trained with quantization in mind, often reducing accuracy loss compared to post-training quantization.

Through this chapter, you have gained the tools necessary to adapt a model's core behavior and specialized knowledge to meet specific requirements. The fine-tuning and prompt engineering techniques ensure that the engine of your AI system is powerful, accurate, and aligned with your goals. However, even the most perfectly tuned model is limited by a fixed cutoff date. The model only knows what it learned during its last training session.

To build truly dynamic systems that can interact with real-time data, private documents, and massive external databases without the need for constant retraining, we must look beyond the model's internal weights. In the next chapter, we will explore Retrieval-Augmented Generation (RAG), a framework that bridges the gap between a model's inherent reasoning capabilities and the vast, ever-changing world of external information.

CHAPTER 5

Retrieval-Augmented Generation

LLMs are highly capable of processing vast amounts of information and developing a strong general understanding of it. Their training on large and diverse datasets enables them to answer a wide range of questions across many topics. However, this knowledge is limited to what is contained in their training data. This means they may struggle to answer certain questions about unfamiliar topics or produce responses that do not align with a user's specific goals. This limitation is expected since the model's knowledge and biases are shaped by its training data.

While such gaps may be acceptable in general-purpose applications, they become critical in enterprise applications where responses must be based on the company's proprietary data, must be accurate, and must be up to date, leaving no room for hallucination. Because enterprise data can change frequently, retraining an LLM every time the data is updated is impractical, especially when only a small portion of the information changes.

This chapter discusses how this challenge can be addressed through retrieval-augmented generation (RAG), an approach that enhances the LLM's context with relevant and up-to-date knowledge from external data sources. By integrating retrieval into the generation process, RAG allows LLMs to produce accurate, up-to-date, and context-specific answers without the need for continual retraining. The chapter concludes by introducing modern variations of the standard RAG model.

5.1 Introduction

LLMs trained on general data typically lack deep knowledge of specialized domains. For example, an LLM can provide general advice on customer service best practices such as being polite and empathetic. However, if asked for specific troubleshooting steps for a

© Ahmed Fawzy Gad 2026
A. F. Gad, *Transformers and Large Language Models*, https://doi.org/10.1007/979-8-8688-2785-3_5

proprietary software product, it may struggle to give accurate answers unless it has been trained on internal company documentation or technical support data.

Moreover, an LLM can provide broad medical advice but lack the precision needed for accurate diagnosis. For example, an LLM can explain common symptoms of diabetes, such as increased thirst and fatigue. But if asked to analyze a patient's lab results and provide a definitive diagnosis, it may fail because it lacks access to patient history, medical imaging, and specialized clinical guidelines used by doctors. Without training on specific medical datasets, it cannot replace a licensed healthcare professional.

LLMs are clever in doing jobs that do not depend on up-to-date information, like text summarization, rephrasing, correcting grammar mistakes, etc. Whenever the task depends on domain-specific and up-to-date information, it struggles unless it is given external support.

Although a generic LLM can answer prompts across many domains, several limitations remain:

1. **Lack of Source**: LLMs typically do not provide the origin of the information used to generate a response. Without citations, users may find it difficult to verify and trust the output.

2. **Outdated Knowledge**: The information stored in an LLM's parameters reflects the data available at the time of training. Since many facts change over time (e.g., the average income of citizens), a model that has not been updated may produce outdated or inaccurate answers.

3. **Limited Domain Specialization**: While a general-purpose LLM can offer broad responses across fields such as medicine, law, or engineering, it often lacks the precision and depth required for expert-level, domain-specific tasks.

Asking an LLM a question is like seeking an answer from someone's mind. It provides knowledge, but without a trusted source or reference, its reliability remains uncertain. So, you would be better off asking the person to give a reference to prove the information given. This can be done by visiting a library and referencing books with information proving whatever has been claimed. This cuts any doubts about the reliability and trust of the information given.

Training LLMs is a complex and resource-intensive process that presents several challenges:

1. **Model Scale**: Modern LLMs often contain billions of parameters, requiring high-performance computing infrastructure that can cost millions of dollars.

2. **Data Requirements**: The large models must be trained on massive datasets containing trillions of words to achieve competitive performance.

3. **Evaluation Difficulty**: Assessing model quality is not always straightforward and often involves subjective, human-guided methods such as RLHF.

Due to these challenges, it is uncommon for most organizations to train an LLM entirely from scratch on proprietary data. Only institutions with enough funding, technical expertise, and computational resources can feasibly train an LLM.

Retrieval-augmented generation (RAG) is a technique that enables an LLM to incorporate external, domain-specific information without the need for additional training or fine-tuning. In this approach, the model is provided with retrieved content (e.g., documents, knowledge base entries, or database records) as contextual input. This allows the LLM to generate responses that include information it was not exposed to during its original training.

The retrieved information is not embedded into the model's parameters. Instead, the model uses it only during the generation process to respond to a specific prompt. Once the interaction ends, the model retains no knowledge of this information.

Because LLMs are trained on static datasets, their knowledge quickly becomes outdated. RAG has therefore become an essential component of many modern generative AI applications, including OpenAI's *ChatGPT*, Anthropic's *Claude*, and Google's *Gemini*. When presented with a question beyond the model's internal knowledge, the system can perform an external search to retrieve relevant context. This is usually by querying the web. This retrieved information is then supplied to the LLM as external context, enabling it to generate a response that reflects the most current information available.

5.2 RAG Types

The RAG architecture can be different based on the embedding technique:

1.   Bi-encoder

2.   Cross-encoder

5.2.1 RAG with Bi-Encoder

As the name suggests, bi-encoder RAG uses the same encoder (embedding model) independently to generate vector representations for both the input query and each external text chunk. This means that for each prompt–chunk pair, two separate embedding vectors are produced. The first one is for the prompt and the second one for the chunk. Figure 5-1 summarizes its workflow, which consists of these steps:

1.   Chunking

2.   Embedding

3.   Storage

4.   Retrieval

5.   Generation

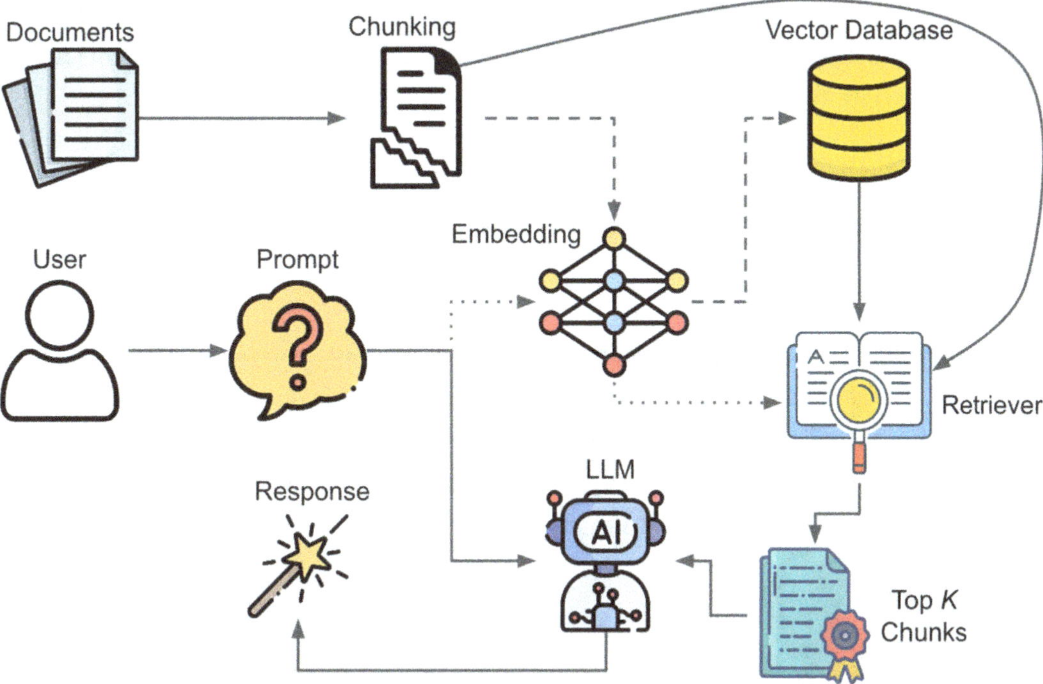

Figure 5-1. *Bi-encoder RAG workflow. The external documents are converted into chunks of text that are embedded, where the embedding vectors are stored into a vector database. When the user submits a prompt that needs external sources for answering, the prompt is embedded using the same embedding model. The embedded prompt is used to retrieve relevant vectors from the database. The text chunks are then passed to the LLM alongside the prompt to generate a response*

Chunking and Embedding

External documents are first divided into smaller semantic segments, or *chunks*, each focused on a single topic. Chunking is the process of breaking down large documents into smaller, manageable pieces. In a RAG system, we do this because LLM models have a limited context window, and they can only read a certain amount of text at once.

Think of it like an extra-large pizza. Normal people can't eat an entire 45 cm extra-large pizza in one single bite, unless they are a woodchipper. Instead, people slice it into equal pieces so it's easier to handle and consume. In a RAG system, we do this because AI models have a limited context window, and they can only swallow a certain amount of information at once. We try to split the text at natural points (like paragraphs or sentences) so the meaning stays intact. However, it is recommended that each chunk maintain thematic consistency.

Each chunk is then processed by an embedding model to produce a vector representation that captures its semantic meaning. Chunk size is determined by a maximum token limit to ensure compatibility with embedding and model input constraints.

The dimensionality of embedding vectors plays a significant role in the retrieval speed of a vector database. Higher-dimensional vectors increase computational cost and latency during similarity search. One approach to improve efficiency is dimensionality reduction, which projects the original high-dimensional vectors into a lower-dimensional space while preserving their relative similarity. This is typically achieved using a projection matrix or other transformation techniques.

Storage

All embeddings are stored in a vector database optimized for fast and efficient similarity search. Examples of the vector database engines are Chroma and Pinecone. Such a kind of database is optimized for indexing the information by quickly searching the database by similarity. One algorithm that can be applied is the nearest neighbor.

Traditional relational databases like SQL are efficient at indexing structured data. They can retrieve data with exact matches to the query. RAG retrieves the matches using similarity metrics without necessarily having exact matches using algorithms like approximate nearest neighbors (ANN). RAG also uses high-dimensional vectors, which are bottlenecks for traditional SQL databases. With more vectors inserted, the speed of retrieving the matches using SQL is another bottleneck. This is why vector databases are preferred over relational databases.

For bi-encoding RAG, the vector database is created only once. It is only updated when the documents are updated to have the most up-to-date information fed to the generative LLM.

Retrieval

When a user submits a prompt, it is embedded using the same model, and the retriever searches the database for the most relevant chunks. The semantic similarity scores between the embedded prompt and the entries in the vector database are calculated. The cosine similarity is an example of the similarity metrics.

Traditionally, the query/prompt embedding vector must be compared to each vector in the database. But this comes against speed. This is why some clustering algorithms (e.g., K-Means) are applied to cluster the vectors based on their similarity using a distance metric such as cosine similarity. This way, the search scope is narrowed down to just a few groups of vectors instead of all the vectors.

Instead of using the embedding vectors' similarity, reranking methods (e.g., BERT-based rankers) can also be applied to measure the contextual relevance so that the decision is more accurate when rejecting a retrieved chunk.

To further improve relevance, redundancy can be minimized. Some retrieved chunks may contain overlapping or identical information, which can be detected by measuring pairwise similarity among the retrieved K chunks. Chunks with a high similarity score (e.g., $\geq 90\%$ cosine similarity) can be excluded from the final set. Similarly, deduplication can also be applied during the embedding stage to prevent adding new chunks that are nearly identical to existing ones in the database.

Generation

At the generation step, the LLM receives both the user's textual prompt and the K most relevant text chunks as additional context. Using this combined input, the model generates a response that has domain-specific or private information without requiring retraining or fine-tuning. This approach allows the LLM to provide accurate, domain-specific, and up-to-date information answers while preserving its original general-purpose capabilities.

> *In a RAG system, the LLM receives the original text of the retrieved chunks, not their embedding vectors.*

Metadata

It is important to note that, in addition to storing the embedding vectors, the system also saves metadata associated with each chunk. This metadata may include a reference to the text chunk and the source file name. Such metadata plays a critical role in downstream operations as it allows the retriever to filter results based on attributes (e.g., only return chunks from a specific document) and facilitates traceability by linking a retrieved chunk back to its original source.

Some extensions might be added to relational databases to help them support vector search capabilities. An example is the pgvector extension for the PostgreSQL database.

Chunk Size

In a bi-encoder RAG architecture, the source text is divided into semantic chunks, with each chunk converted into an embedding vector and stored in a vector database. The chunk size is an important design parameter that directly affects the efficiency and performance of the RAG system.

Some models use token-based embeddings where each word in the chunk is embedded. Then a single vector for the chunk is created out of all the token embedding vectors by a pooling operation. Others may embed sentences or the complete chunk directly as a single embedding vector.

Smaller chunks increase semantic precision but result in a larger number of entries in the vector database. This expansion can significantly increase storage requirements and retrieval latency as the search must process more vectors. Conversely, larger chunks reduce the total number of database entries, which improves the search speed and lowers storage needs. However, a single vector may then represent multiple, potentially unrelated topics within the same chunk, reducing retrieval accuracy.

It is even possible to create a single embedding vector for the entire document. But it is more efficient to use embedding vectors out of small semantic chunks of text for some reasons, like:

1. Focus on the details.

2. Precise indexing.

For example, a single document might have sections discussing different topics like gas prices, inflation, and supply chains even if they are under the same umbrella of economics. Embedding more text into the same embedding vector makes it hard to focus on the different semantics of the different sections of the document. It is better to create embedding vectors for small chunks of text to keep focusing on the details.

When indexing the database, the model can reference the exact passage or chunk relevant to the prompt when the smaller chunks are used. If an entire large document is embedded, the results will not be precise, as the RAG will retrieve the large document instead of focusing on the exact relevant chunk.

But be aware that using too small a chunk size may be out of the main context. For example, a document might be discussing the limitations of dispensing some drugs. One small paragraph discusses the number of people diagnosed with diabetes. Later, this paragraph is linked to the context of drugs not being available to the market. If the chunk size is small, the small paragraph about diabetes may be used alone, which is off the main context. This is why some embedding-based chunking techniques merge 2 chunks together if their embedding vectors are very similar.

Numeric Vectors

The advantage of converting text into a numerical vector representation is that it enables retrieval based on semantic similarity rather than simple word matching. This allows the system to identify text that is conceptually related to the user's query, even if different words or phrases are used.

Without embedding, retrieval often relies on exact keyword matching. For example, if a user submits the following prompt:

Prompt: Who is the best football player in Africa?

Then a keyword-based search would only return results containing the same words: *football*, *player*, and *Africa*. However, the external documents may have semantically relevant information that is only expressed using different terminology, such as *soccer* instead of *football*, or *dark continent* instead of *Africa*.

Text: The top soccer players in the dark continent ...

An embedding-based search can recognize these relationships in meaning, retrieving the most relevant content even when there is no direct word match.

Example

This section provides an example of building a bi-encoder RAG system using the *LlamaIndex* framework, which is designed to support the development of context-augmented LLM applications.

Since we will be using the OpenAI API, the first step is to set the API key as an environment variable named OPENAI_API_KEY.

```python
import os

os.environ["OPENAI_API_KEY"] = "sk-..."
```

The LlamaIndex library allows you to specify the OpenAI embedding model used to generate embedding vectors for both the chunks and the prompt through the OpenAIEmbedding class. In the same way, the OpenAI generative model responsible for producing the final response can be set using the OpenAI class. In our example, we use the text-embedding-3-large model for embeddings and the gpt-4o model for generation.

```python
from llama_index.llms.openai import OpenAI
from llama_index.embeddings.openai import OpenAIEmbedding
from llama_index.core import Settings

Settings.embed_model = OpenAIEmbedding(model="text-embedding-3-large")

Settings.llm = OpenAI(model="gpt-4o")
```

Next, we specify the directory containing the documents using the SimpleDirectoryReader class. Each document is automatically split into smaller chunks, which are then prepared for embedding. Every chunk is represented as an instance of the llama_index.core.schema.Document class and assigned a unique identifier.

```python
from llama_index.core import SimpleDirectoryReader

documents = SimpleDirectoryReader("/Users/ahmedgad/Desktop/Data").
load_data()
```

Each instance of the Document class stores metadata about its corresponding chunk, such as:

1. File name

2. File type

3. Page number

4. Document ID

5. Document creation date

Once the documents are loaded, an index is created in which every chunk is embedded using the specified embedding model. The resulting embedding vectors are then stored in an in-memory vector database to enable efficient semantic search and retrieval during query time.

```python
from llama_index.core import VectorStoreIndex

index = VectorStoreIndex.from_documents(documents)
```

Once the index is created, the next step is to create a query engine, which will later be used to accept user queries and call the generative model to produce responses. The query engine is responsible for retrieving the top K most relevant chunks from the vector database and passing them to the LLM. The parameter similarity_top_k can be used to control how many chunks are retrieved for each query.

```python
query_engine = index.as_query_engine(similarity_top_k=5)
```

Finally, we can issue a query using the query() method to test the bi-encoder RAG. For example, if the dataset contains a CV in PDF format, we can ask about the candidate's name. Since one of the chunks will include the candidate's name, the generative LLM will be able to provide the correct answer.

```python
response = query_engine.query("What is the candidate's name?")
```

The chunks that are retrieved and passed into the model's context can be inspected using the source_nodes property of the response object. From these source nodes, you can also access additional metadata such as the similarity score between the query and each chunk as well as the actual chunk text content.

```python
for i, source_node in enumerate(response.source_nodes):
    print(f"\nContext Chunk Index: {i}")
    print(f"Score: {source_node.score}")
    print(f"Doc ID: {source_node.node.ref_doc_id}")
    print(f"File: {source_node.metadata.get('file_path')}")
    print(f"Start char: {source_node.node.start_char_idx}, End char: 
{source_node.node.end_char_idx}")
    print(source_node.get_content())
```

5.2.2 RAG with Cross-Encoder

Bi-encoder RAG has the limitation of generating independent embedding vectors for the text chunk and the prompt without any direct interaction between them. This leads to some drawbacks, including:

1. Information loss can occur because each text chunk or prompt is compressed into a single embedding vector, which may not fully preserve all details.

2. Loss of contextual interactions between the prompt and the chunk since they are encoded separately, and their relationship is not modeled during embedding generation.

To highlight the limitation of bi-encoding, assume there are two text documents, where one is about the apple as a fruit and another as the Apple company. If the prompt is, for example:

Give insights about apple and orange prices next week.

It is obvious that the prompt is about fruits since the words apple and orange appear together.

When the prompt is fed to the RAG model, then it retrieves chunks from the two documents about the fruit and the company as relevant despite being different in context. Since the word *apple* appears in both contexts (fruit and company) and without the query–chunk interaction, the embedding space may place them close together just because of shared tokens and general semantic similarity.

Fruit: The price of apple is going to increase in grocery stores next week.

Company: The price of the new released apple device is going to reduce next week.

The model is not aware of the context when the text chunks and the prompt are independently embedded into separate vectors. As a result, the retriever may return chunks from both documents even though only one context (fruit price vs. device price) is truly relevant.

The limitations of bi-encoding are addressed by the cross-encoder approach, as illustrated in Figure 5-2. Instead of generating separate embeddings for the prompt and the chunk, both are fed into a ranker model simultaneously.

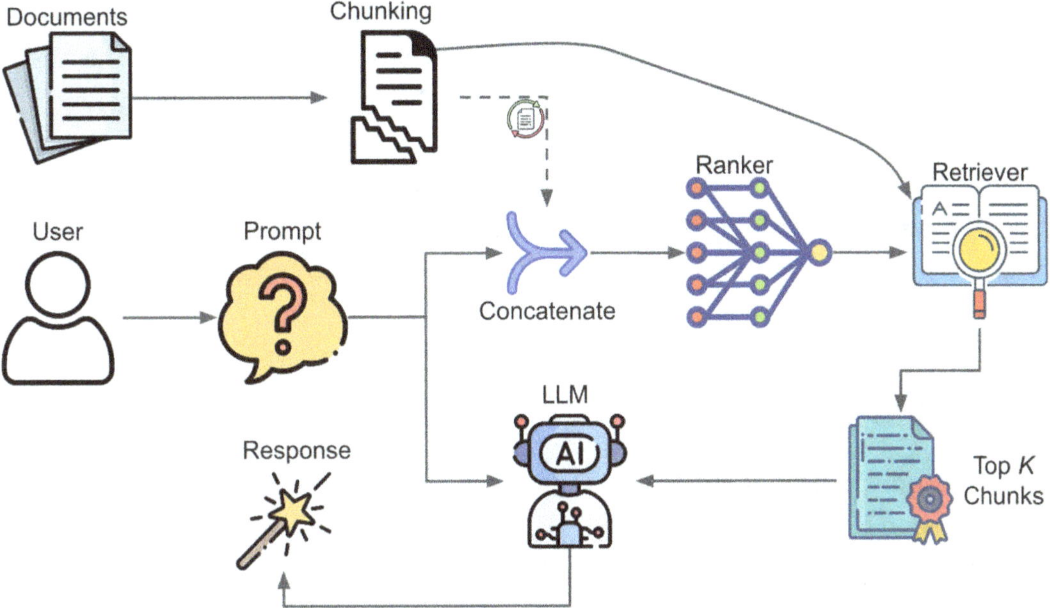

Figure 5-2. *Workflow of the cross-encoder RAG. Each chunk in the dataset is concatenated with the prompt and fed into a ranker model to produce a similarity score for the prompt–chunk pair. The top-K chunks with the highest scores are then retrieved and provided to the LLM as an external context for generating the final response to the prompt*

The ranker model consists of two components:

1. Encoder

2. Classifier

The encoder processes the prompt and the chunk together to capture fine-grained interactions between individual tokens in both texts. Since relevance scores are computed on the fly, there is no need to store embeddings in an external vector database.

The encoder's output is passed to a classification layer that produces a relevance score indicating the similarity between the prompt and each chunk. Based on these scores, the top K chunks across all documents are selected and, together with the prompt, provided as context to the generative model to produce the final response.

The cross-encoder RAG retrieves more relevant chunks due to its ability to capture relationships between individual tokens in the prompt and each chunk. However, this comes at the cost of significantly higher computational time. For long documents containing many paragraphs or passages, the cross-encoder must compute embeddings and similarity scores between each prompt and each paragraph individually. This process is not easily scalable and is generally unsuitable for real-time applications.

In contrast, bi-encoding is much faster because the embeddings for all text chunks in the database are computed once and stored in a vector database. At inference time, only the prompt's embedding is computed. With efficient indexing, the retriever can quickly locate and return the most relevant context despite the fact that this comes at the expense of some accuracy.

5.2.3 Hybrid

Both bi-encoder and cross-encoder RAG models have their pros and cons. As summarized in Figure 5-3, the bi-encoder RAG is faster because each chunk is embedded only once and stored for reuse. But it is generally less accurate, as it embeds the chunks independently from the prompt. The cross-encoder achieves higher accuracy by capturing token-level interactions between the prompt and each chunk. But this comes at the cost of slower retrieval because it must measure the similarity score between all chunks and each prompt.

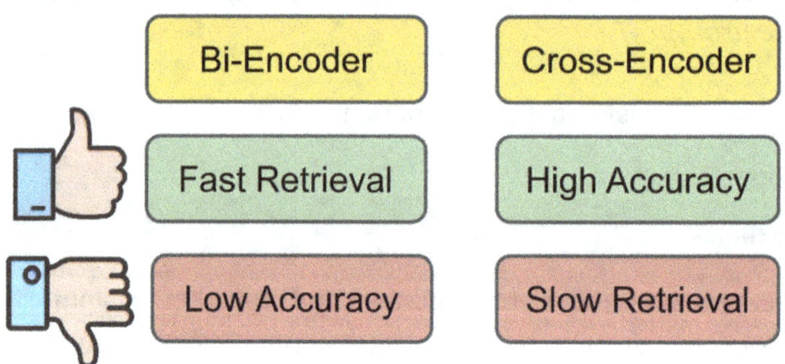

Figure 5-3. *Comparison between bi-encoder and cross-encoder RAG models*

Instead of relying on a single RAG type, a hybrid approach can be used to combine the strengths of both models, as shown in Figure 5-4. In this approach, the bi-encoder method is first used to filter the chunks and produce a shortlist of candidates, for example, the top 50 chunks. This reduces the number of chunks that need to be evaluated by the ranker.

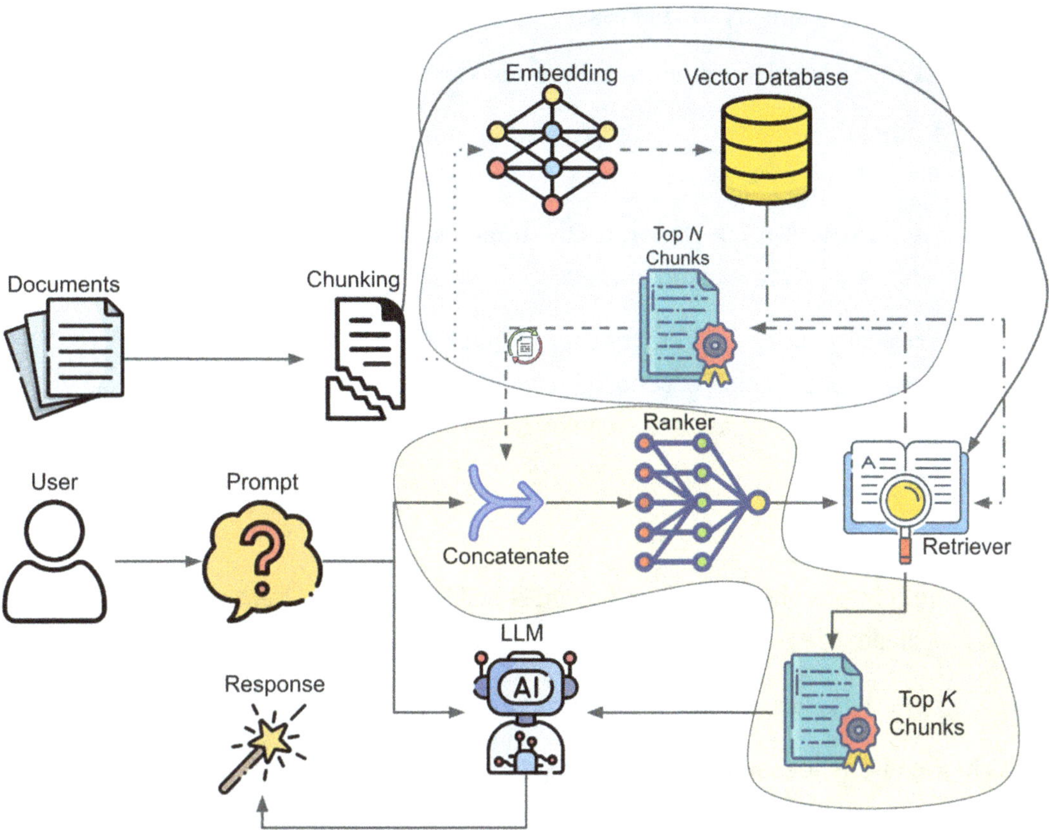

Figure 5-4. *Hybrid RAG architecture combining the advantages of both bi-encoder and cross-encoder methods. A vector database stores the embedding vectors of all chunks. From this database, a shortlist of candidate chunks is retrieved using the bi-encoder. Only these shortlisted chunks are then processed by the cross-encoder ranker, which scores and selects the most relevant ones to be passed to the generative model*

The cross-encoder ranker is then applied to this reduced set to score each prompt–chunk pair based on their relevance. Finally, the top K (e.g., 5) ranked chunks from this list are selected and provided to the generative model as context for producing the answer.

Example

The procedure for building a hybrid RAG system closely mirrors the steps used in constructing a bi-encoder RAG. The key difference lies in the addition of a re-ranking model, which is applied to the retrieved chunks to improve retrieval accuracy.

To begin, we first specify the necessary OpenAI settings:

1. **API Key**: Used to authenticate requests to the OpenAI API.

2. **Embedding Model**: Converts the text chunks and queries into embedding vectors.

3. **Generative Model**: Generates the final response based on the retrieved and re-ranked chunks.

```python
from llama_index.llms.openai import OpenAI
from llama_index.embeddings.openai import OpenAIEmbedding
from llama_index.core import Settings
import os

os.environ["OPENAI_API_KEY"] = "sk-..."

Settings.embed_model = OpenAIEmbedding(model="text-embedding-3-large")

Settings.llm = OpenAI(model="gpt-4o")
```

As in the bi-encoder setup, the next step is to specify the directory that contains the source documents. These documents are automatically parsed and divided into smaller chunks to ensure they fit within the constraints of the embedding model.

Once the chunks are created, an index is built to generate embedding vectors for each chunk using the specified embedding model. These embedding vectors are stored in a vector database to use for semantic search during query time.

```python
from llama_index.core import SimpleDirectoryReader, VectorStoreIndex

documents = SimpleDirectoryReader("/Users/ahmedgad/Desktop/Data").load_data()

index = VectorStoreIndex.from_documents(documents)
```

The distinctive step in the hybrid RAG workflow is the use of a cross-encoder reranker. This reranker is created with the SentenceTransformerRerank class that evaluates the relevance of each retrieved chunk with respect to the query. Unlike the bi-encoder retriever that relies on embedding similarity alone, the cross-encoder reranker processes the query and each chunk together, enabling it to capture fine-grained token-level interactions.

The reranker assigns a relevance score to every candidate chunk and then reorders them according to these scores. To limit the amount of context passed to the generative model, the reranker keeps only the top N chunks as specified through the `top_n` parameter. In our example, only the top 5 chunks are retained, ensuring that the most relevant information is preserved.

```python
from llama_index.core.postprocessor import SentenceTransformerRerank

reranker = SentenceTransformerRerank(model="cross-encoder/ms-marco-MiniLM-L-6-v2", top_n=5)
```

The query engine is constructed from the index using the `as_query_engine()` method. In the hybrid RAG setup, two key arguments are provided:

1. `similarity_top_k`: This parameter specifies how many chunks are retrieved in the initial stage using the bi-encoder retriever. The query is first converted into an embedding vector, and the retriever searches the vector database to fetch the top K chunks whose embeddings are most similar to the query embedding. The similarity is typically measured using a distance metric such as cosine similarity.

2. `node_postprocessors`: After the top K chunks are retrieved, this parameter defines a set of post-processing steps to apply before passing the context to the generative model. In the hybrid approach, it only includes the cross-encoder reranker created earlier where the top N chunks (as defined by the reranker's `top_n` setting) are retained and provided to the LLM as the final context.

```python
query_engine = index.as_query_engine(similarity_top_k=40,
                                      node_postprocessors=[reranker])
```

Finally, the query engine can be invoked by passing the user's query. The engine retrieves the top N most relevant chunks as selected after bi-encoder retrieval and cross-encoder reranking. Then it uses such chunks as context for the generative model. Together with the query, information is provided to the LLM, which then produces the final response.

```python
response = query_engine.query("What is the candidate's name?")
print(response)
```

In addition to generating the final response, we can inspect the top chunks after re-ranking to see their content and their associated similarity scores. This makes it possible to verify which passages were selected as context for the LLM and evaluate the quality of retrieval.

```python
for i, source_node in enumerate(response.source_nodes):
    print(f"\nContext Chunk Index: {i}")
    print(f"Score: {source_node.score}")
    print(source_node.node.get_content())
```

5.3 Multi-Query Prompt

The retrieval step in a RAG system is critical because it selects the most relevant information from text documents or other data sources to provide context for the language model. When the retriever successfully identifies the best chunks, the model can generate a high-quality answer. If the retrieval step fails, the generated response will likely be incomplete or unsatisfactory.

Simple queries in a RAG system usually have a clear intent. This makes the retrieval process straightforward. For example, consider a grocery store that maintains data about its profits. A direct query might be the following:

What is the profit from selling lentils in 2025?

Given that the data is organized by year and by product category, there may be a document containing statistics about pulses. Such a document would include the profits from selling different goods in each quarter of every year at the grocery store. In this case, the RAG model would not struggle to locate the most relevant text chunk to answer the query.

However, the prompt can also be more complex, such as:

Compare the profits from selling lentils and pasta in quarter four of 2025 with those in quarter four of 2024.

The retriever must recognize that the query is asking about four objectives:

1. Two objectives about the profit from lentils in quarter four of 2025 and quarter four of 2024

2. Another two about the profit from pasta in quarter four of 2025 and quarter four of 2024

Unfortunately, a RAG retriever typically processes the query as a whole when searching for matching chunks. This limitation is especially evident in a bi-encoder retriever where the entire prompt is converted into a single embedding vector. That vector is then compared with those stored in the vector database to find the most similar matches. Because no single chunk in the data fully addresses the entire query, the retrieved results are often insufficient to provide a complete answer.

This limitation can be addressed by splitting the original query into several simpler queries, as illustrated in Figure 5-5:

1.   The profit from selling lentils in quarter four of 2025

2.   The profit from selling lentils in quarter four of 2024

3.   The profit from selling pasta in quarter four of 2025

4.   The profit from selling pasta in quarter four of 2024

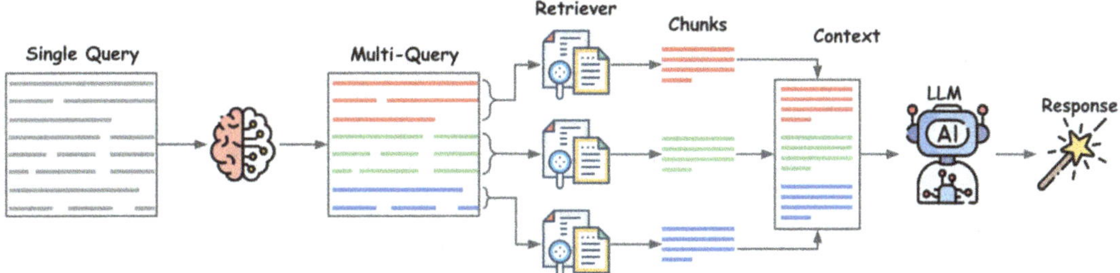

Figure 5-5. *Splitting a single complex query into multiple simpler queries, each with a clear objective. The retrieved chunks from these queries are then combined and used as context for the generative language model to produce a complete response*

Each query is focused on a single objective. Depending on the structure and organization of the data, the query may need to be divided into additional subqueries. For instance, if the dataset contains only monthly profit records rather than quarterly totals, it is more effective to retrieve the profit for each month within quarter four and then use those results as context for the generative model.

To automate this process, a language model can be used to interpret the original query and determine whether it should be divided into multiple subqueries as indicated in Figure 5-5. Each subquery is then passed to the RAG system to retrieve the most relevant chunk. All retrieved chunks are combined to form the context for the generative model.

Compared to traditional RAG models that attempt to process the entire complex query as a single unit, splitting the query into simpler subqueries provides the generative model with more complete and relevant information. As a result, the language model is more likely to generate a higher-quality answer.

5.4 RAG Disadvantages

While RAG provides a fast and flexible method for adapting a language model to specific sources of information without retraining, it also introduces several important limitations, including:

1. Safety constraints

2. Inference latency

3. Dependent on retrieval quality

4. Overhead

5. Adversarial queries

5.4.1 Safety

One limitation stems from the reliance on a pre-trained base model. Such models are typically developed by large organizations and incorporate built-in safety policies and alignment mechanisms. These mechanisms may impose restrictions that prevent the model from generating certain types of answers even when the retrieved context contains the relevant information. In practice, this means that if a user's query conflicts with the model's internal safety rules or policy guidelines, the system may decline to respond or return a filtered answer. This limits its usefulness in some applications.

5.4.2 Latency

Another limitation concerns inference speed. Unlike using the base language model directly, a RAG pipeline must first carry out a retrieval step. This involves searching a document collection, ranking candidate chunks, and selecting the most relevant pieces of text. Only after this retrieval process is completed can the model generate a response. This additional stage introduces latency and can make RAG slower than direct inference with the model alone. This limitation becomes more significant in systems that operate over very large document collections or in applications where real-time responsiveness is essential.

5.4.3 Retrieval Quality

The core job of RAG models is retrieving the relevant information to be fed to the generative model as a context. As a result, the accuracy of RAG models heavily depends on the retriever's ability to select the most relevant chunks. If the retriever fails to capture critical context, the LLM may generate incomplete or misleading answers.

5.4.4 Overhead

Building and maintaining a RAG pipeline requires additional infrastructure such as a retriever and a vector database. The vector database must also be periodically refreshed whenever documents are added, removed, or updated. This introduces operational overhead compared to using a standalone language model. However, despite this added complexity, RAG remains far more efficient than training or retraining a large language model from scratch, as it allows the system to incorporate new information quickly without the need for costly model updates.

5.4.5 Adversarial Queries

RAG is often employed to ensure that a system can generate responses grounded in the most up-to-date information available. In many cases, the underlying sources are private or sensitive and are not intended to be exposed directly to the end user. The model should therefore use these sources only to generate an appropriate answer without revealing the underlying documents themselves because they may contain more information than the user is authorized to access.

For example, consider a customer service setting where an employee is permitted to view a user's subscription status but not the user's billing details. When given a prompt, the RAG retriever may gather the entire customer profile (including both subscription and billing information) and provide it as context to the generative model. The model is expected to generate a response that reveals only the relevant subscription details while suppressing sensitive information such as credit card numbers.

However, this setup introduces a potential security risk. Through prompt injection or prompt manipulation, a malicious user might include hidden instructions within their query to trick the generative model into revealing the retrieved context. In the earlier example, an employee could craft a prompt that causes the model to output the entire retrieved customer profile to expose billing details that should remain confidential.

To address some of these issues, such as latency and overhead, fine-tuning could be combined with RAG. Fine-tuning allows the base model to adapt to a specific domain such as finance, law, or medicine. This makes the model more familiar with specialized terminology, concepts, and reasoning patterns. This increases the model's reliability and reduces dependence on safety filters that may not align with domain requirements.

Once fine-tuned, RAG can be layered on top to provide the model with the most recent and contextually relevant information. This ensures that responses are both domain-specific and up to date. In this hybrid approach, fine-tuning enhances the model's expertise while RAG ensures adaptability to changing knowledge sources. Together, they create a more effective system that balances accuracy, domain alignment, and responsiveness.

5.5 Retrieval Evaluation

The RAG model involves several key stages (chunking, retrieval, and generation), each of which is critical to the quality of the final response produced by the language model. To ensure accuracy and relevance, it is important to evaluate not only the generated response but also the intermediate retrieval process because the quality of retrieved chunks directly influences the final output.

Commonly used metrics for evaluating the relevance of retrieved information include:

1. Precision

2. Recall

3. Mean reciprocal rank

It is important to note that these metrics are computed over the top K retrieved chunks and require the availability of ground-truth data. Such data is typically constructed by mapping a set of queries or prompts to their corresponding relevant chunks. This ground truth serves as the reference against which retrieval performance is evaluated.

5.5.1 Confusion Matrix

RAG models can be evaluated using precision and recall. Since these metrics are derived from the confusion matrix, it is useful to first review this concept. The confusion matrix provides a structured summary of how well a classification model distinguishes between different classes by indicating both correct and incorrect classifications. In the context of RAG, the retrieval process can be framed as a binary classification problem in which each retrieved chunk is classified as either relevant or irrelevant to the query.

Although the confusion matrix can be applied to both binary and multi-class classification problems, our focus will be on the binary case where the two classes are typically referred to as *positive* and *negative.* For RAG, a relevant chunk can be considered positive, while an irrelevant chunk is negative.

The confusion matrix for a binary classification problem consists of four entries, as illustrated in Figure 5-6:

1. **True Positive (TP)**: A positive sample that is correctly classified as positive.

2. **False Positive (FP)**: A negative sample that is incorrectly classified as positive.

3. **False Negative (FN)**: A positive sample that is incorrectly classified as negative.

4. **True Negative (TN)**: A negative sample that is correctly classified as negative.

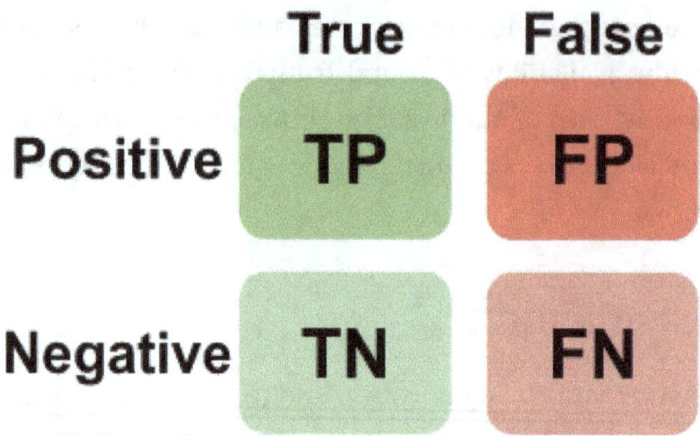

Figure 5-6. *The confusion matrix is a tool for evaluating the performance of classification models. One dimension represents the actual class labels (e.g., positive and negative), while the other dimension represents the predicted class labels. The intersection of these dimensions captures whether each prediction is correct or incorrect*

When the term begins with *False*, it indicates an incorrect classification compared to *True* which denotes a correct classification. The second part of the term refers to the class to which the sample was assigned. For example, a *True Negative* refers to a sample that is correctly classified as belonging to the negative class.

With ground-truth data available, the correct class for each sample is already known. The model's predictions can then be directly compared against these ground-truth labels to compute the evaluation metrics such as precision and recall.

Precision

Precision represents the proportion of retrieved chunks that are actually relevant relative to all chunks classified as relevant, whether correctly (True Positives) or incorrectly (False Positives).

$$Precision = \frac{TP}{TP + FP} \tag{1}$$

It measures the accuracy of retrieving the relevant chunks. A higher precision indicates that the retriever is effective at returning the relevant chunks. In this case, the number of true positives increases while the number of false positives (i.e., chunks incorrectly classified as relevant) decreases. This results in an overall increase in precision.

Recall

Recall represents the proportion of relevant chunks successfully retrieved out of the total number of relevant chunks available.

$$Recall = \frac{TP}{TP + FN} \tag{2}$$

It measures the ability of the retriever to capture all relevant chunks. A higher recall indicates that fewer relevant chunks were missed. As the retriever returns more of the truly relevant chunks, the number of false negatives (i.e., relevant chunks that were incorrectly classified as irrelevant) decreases. This increases the recall.

5.5.2 Mean Reciprocal Rank (MRR)

Mean Reciprocal Rank (MRR) is a metric used to evaluate the effectiveness of a retriever in ranking relevant documents among the top K retrieved results. Each retrieved chunk is assigned a rank as a positive integer between 1 and K, where rank 1 indicates the most relevant chunk. As the rank number increases, the relevance of the chunk decreases. Ideally, the most relevant chunk should appear at rank 1.

MRR measures how early the first relevant chunk appears in the ranked list by computing the reciprocal of its rank. The overall MRR is then calculated as the average of these reciprocal ranks across all queries:

$$MRR = \frac{1}{N}\sum_{i=1}^{N}\frac{1}{rank_i} \tag{3}$$

Where N is the total number of queries and $\mathbf{rank}_i$ is the position of the first relevant chunk for the i-th query.

Since ranks are expressed as positive integers, MRR reaches its maximum value when $\mathbf{rank}_i = 1$. This is because the reciprocal $1/\mathbf{rank}_i$ is largest (1) when the rank is 1. As $\mathbf{rank}_i$ increases, the reciprocal decreases. MRR achieves its maximum when the most relevant chunk is consistently ranked first across all retrievals.

5.6 Generation Evaluation

In addition to evaluating the retrieval stage, it is also important to assess the generated response produced by the language model. This ensures that the final output is both relevant and correct. Commonly used evaluation methods include:

1. **Human Evaluation:** Human annotators assess the generated responses along dimensions such as correctness, relevance, fluency, and coherence. Although subjective and costly, human evaluation often provides the most reliable benchmark.

2. **Semantic Similarity:** An automatic way that computes the semantic similarity between the generated response and reference answers.

3. **RAG Assessment (RAGAS):** A framework specifically designed for evaluating RAG models. It integrates multiple evaluation dimensions (e.g., faithfulness and answer relevance) to provide a comprehensive measure of quality.

5.6.1 Semantic Similarity

As in traditional supervised learning, evaluation often relies on annotated datasets where each input is paired with an optimal output prepared in advance. In many cases, human annotators are required to create or refine these samples by either generating labels directly or by reviewing and adjusting labels produced by automated tools. When working with LLMs, a key challenge is that ground-truth data is often unavailable or difficult to establish.

For RAG, the model is given a set of queries, and its generated responses are compared against corresponding reference responses. Several metrics can be used, such as:

1. **BLEU (Bilingual Evaluation Understudy):** Measures the overlap between the generated response and the reference response by counting exact word matches. While useful, BLEU is limited because it does not account for semantic similarity, as different words with similar meanings are treated as mismatches.

2. **BERTScore**: Improves on BLEU by measuring semantic similarity between the generated and reference responses. Using contextual embeddings from pretrained models such as BERT or RoBERTa, BERTScore can recognize semantically equivalent words (e.g., *car* and *vehicle*) as matches. This makes it better suited for evaluating natural language generation.

The next code sample uses the `bert-score` library (https://github.com/Tiiiger/bert_score) for calculating the BertScore using Meta's `roberta-large` model. If not installed, use this command to install the library:

```
pip install bert-score
```

The reference text and the generated text are both passed to the model, which computes similarity based on their contextual embeddings.

```python
from bert_score import score

candidate = ["The cat is on the mat."]
reference = ["A cat is sitting on a mat."]

P, R, F1 = score(cands=candidate,
                 refs=reference,
                 model_type="roberta-large",
                 lang="en")

print(f"Precision: {P.item()}\nRecall: {R.item()}\nF1: {F1.item()}")
```

The `score()` function in the `bert_score` package accepts the following key arguments:

1. `cands`: A list of candidate sentences generated by the model.

2. `refs`: A list of the corresponding reference sentences.

3. `model_type`: The pretrained model to use for computing BERTScore (e.g., `"roberta-large"`).

4. `lang`: The language of the input sentences (e.g., `"en"` for English).

Either `model_type` or `lang` must be specified. If only `lang` is provided, a default pretrained model appropriate for that language will be automatically selected.

The `score()` function returns three evaluation metrics:

1. Precision

2. Recall

3. F1

In this example, all three values are above 95%, indicating a very high degree of similarity between the candidate sentence and the reference sentence. High precision suggests that most of the retrieved semantic matches are correct, high recall indicates that most relevant matches were captured, and the high F1 score reflects a strong overall balance between the two.

```
Precision: 0.9629589319229126
Recall: 0.9506745338439941
F1: 0.9567772746086121
```

To visualize the similarity between the tokens in the candidate and reference sentences, we can use the `plot_example()` function. This function accepts parameters similar to those of `score()`, but with one important difference: instead of lists of sentences, it requires a single candidate sentence and a single reference sentence as strings.

```
from bert_score import plot_example

plot_example(candidate=candidate[0],
             reference=reference[0],
             model_type="roberta-large",
             lang="en")
```

The output is a heatmap showing pairwise token-level similarities as illustrated in Figure 5-7. This visualization highlights how closely individual tokens in the candidate sentence align with those in the reference sentence. Candidate tokens are represented along the vertical (Y) axis, while reference tokens are placed along the horizontal (X) axis.

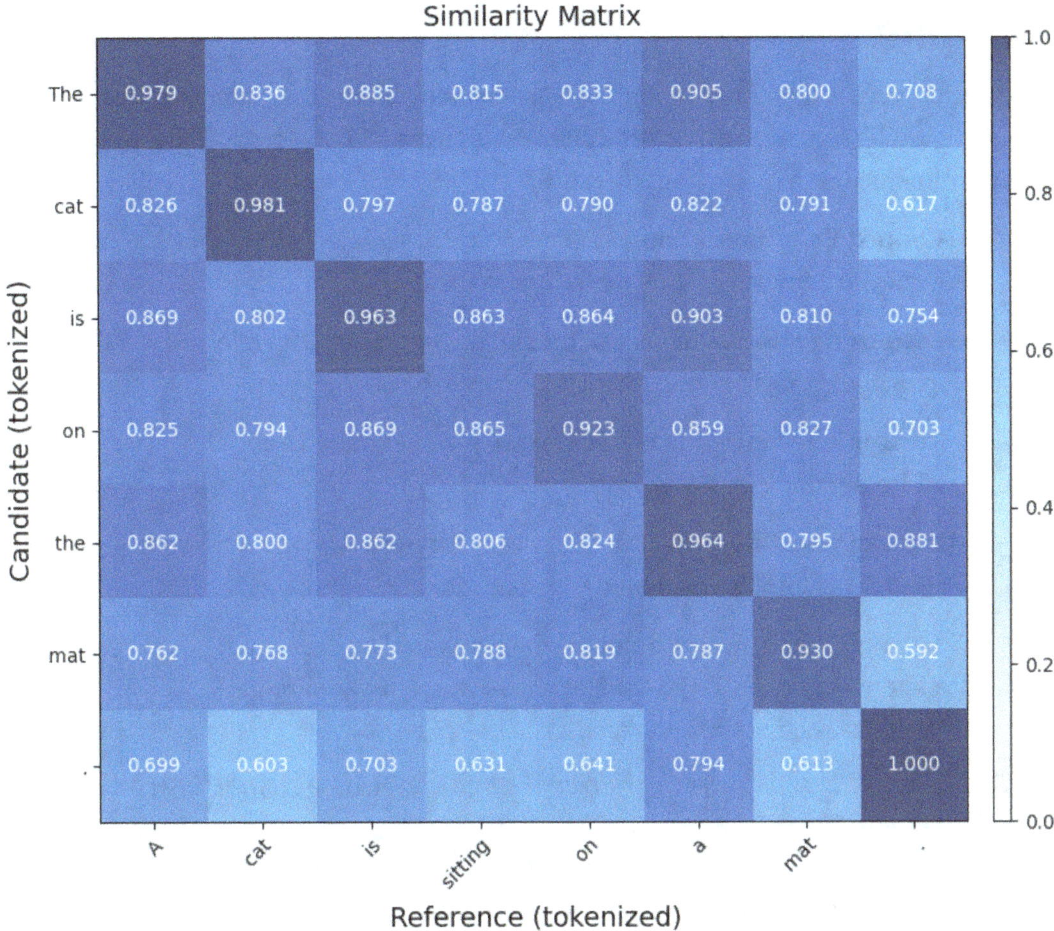

Figure 5-7. Semantic similarity scores between tokens in the candidate response and the reference text. Higher scores indicate greater similarity between the corresponding tokens

For each token in the candidate sentence, the cosine similarity score is computed with respect to the embedding vectors of all tokens in the reference sentence. The result is a matrix of semantic similarity values ranging from 0.0 to 1.0. For example, the token shows high similarity scores with the tokens *sitting* and *on*, reflecting their semantic correlation in context.

5.6.2 RAGAS

RAGAS (Retrieval-Augmented Generation Assessment) is an automated approach for evaluating RAG-generated responses. RAGAS uses expert LLMs (e.g., GPT-4) to score multiple dimensions of quality, such as:

1. **Context Relevance (Context Precision)**: Assesses whether the retrieved context is actually relevant to the query.

2. **Context Recall**: Evaluates whether all the necessary and relevant information has been retrieved to answer the query.

3. **Answer Relevance**: Measures how well the generated answer addresses the user's query.

4. **Faithfulness**: Checks that the answer is supported by the retrieved context (i.e., not hallucinated).

5. **Answer Correctness**: Compares the generated answer against a known reference or ground-truth answer to make sure the response is objectively correct.

RAGAS can also evaluate the retrieval stage of a RAG system using context-related metrics.

RAGAS returns a score between 0 and 1 for each metric or dimension. We can aggregate all of the returned scores to have a compact indicator of overall RAG quality.

Example

To run a basic RAGAS evaluation, first install the required packages. The `ragas` package provides the evaluation framework, while `langchain-openai` enables the use of OpenAI models for both embedding and evaluation.

```
pip3 install ragas
pip3 install langchain-openai
```

As usual, we have to load the OpenAI generative model in addition to an embedding model. As in the next code, we will use GPT-4.1 for text generation and `text-embedding-3-large` for embedding. Remember to set the OpenAI API key to gain access.

In RAGAS, generative models are used for metrics like faithfulness and answer correctness, while embedding models are used for metrics such as context precision, context recall, and answer relevancy.

```python
from langchain_openai import ChatOpenAI, OpenAIEmbeddings

import os

os.environ["OPENAI_API_KEY"] = "sk-..."

llm = ChatOpenAI(model="gpt-4.1")

embeddings = OpenAIEmbeddings(model="text-embedding-3-large")
```

After loading the models, the next step is to prepare the dataset that will be used for RAG evaluation. The dataset typically follows a structured format where each entry contains:

1. user_input

2. retrieved_contexts

3. response

4. reference

```python
from datasets import import Dataset

data = {"user_input": ["What is the capital of Egypt?"],
        "retrieved_contexts": [["Cairo is the capital of Egypt.",
                                "Egypt's capital is very crowded."]],
        "response": ["The capital of Egypt is Cairo."],
        "reference": ["Cairo"]}

dataset = Dataset.from_dict(data)
```

In this example, we will use a single sample to demonstrate the RAG evaluation process.

What is the capital of Egypt?

The retriever is assumed to have returned two chunks of text as a supporting context. These are stored under the `retrieved_contexts` key:

1. *Cairo is the capital of Egypt.*

2. *Egypt's capital is very crowded.*

Based on the retrieved contexts, the language model generates an answer, which is saved in the `response` key:

The capital of Egypt is Cairo.

The `reference` key contains the ground-truth answer against which the model's response is evaluated:

Cairo

In this illustrative example, the context and response are hardcoded for simplicity. However, in a real-world evaluation pipeline, the contexts would be retrieved by the RAG system, and the response would be generated by the language model using those contexts.

Once the models and dataset are prepared, the next step is to build the RAGAS evaluator using the `evaluate()` function. These parameters are passed:

- `dataset`: This is the only required parameter, which contains the queries, retrieved contexts, responses, and references.

- `metrics`: The `metrics` parameter is a list of evaluation metrics. If omitted, RAGAS will automatically select a default set. Check RAGAS documentation (`https://docs.ragas.io/en/stable/concepts/metrics/available_metrics`) for the full list of metrics. In this example, we include:

  1. Context Precision

  2. Context Recall

  3. Answer Relevancy

  4. Faithfulness

  5. Answer Correctness

- llm: Specifies the large language model used for metrics that require generative reasoning, such as *faithfulness* and *answer correctness*. If not provided, RAGAS uses a default LLM.

- embeddings: Specifies the embedding model used for metrics that rely on vector similarity, such as *context precision, context recall,* and *answer relevancy.* If not provided, RAGAS uses a default embedding model.

```python
from ragas import evaluate
from ragas.metrics import context_precision, answer_relevancy,
faithfulness, context_recall, answer_correctness

metrics = [context_precision, context_recall, answer_relevancy,
faithfulness, answer_correctness]
result = evaluate(dataset=dataset,
                  metrics=metrics,
                  llm=llm,
                  embeddings=embeddings)
```

The result returned from the evaluate() function is an object from the class ragas.dataset_schema.EvaluationResult. Printing it returns the score for the used metrics.

```
{'context_precision': 1.0000,
 'context_recall': 1.0000,
 'answer_relevancy': 1.0000,
 'faithfulness': 1.0000,
 'answer_correctness': 0.8956}
```

You can access the individual metric scores (along with the input and output fields) by converting the evaluation result into a Pandas DataFrame. Each row corresponds to one evaluation sample, and it consists of both the raw data and the computed metrics.

```python
df = result.to_pandas()
print(df.keys())
for key in df.keys():
    print(f"{key}: {df[key][0]}")
```

The columns that appear in the DataFrame depend on the fields defined in the dataset (e.g., whether a `reference` answer exists) and the specific metrics selected for evaluation. In our example, the DataFrame includes the following columns:

- `user_input`: The query provided by the user.

- `retrieved_contexts`: The contexts retrieved by the RAG system.

- `response`: The answer generated by the language model.

- `reference`: The ground-truth answer.

- `context_precision`: Context relevance score.

- `context_recall`: Retrieval completeness.

- `answer_relevancy`: The score for how well the answer addresses the query.

- `faithfulness`: The score of the answer's consistency with the retrieved context.

- `answer_correctness` : The score for factual correctness against the reference.

As indicated in Figure 5-8, the `reference` answer must be included in the dataset in order to compute these metrics:

1. `context_precision`

2. `context_recall`

3. `answer_correctness`

4. `answer_relevancy`

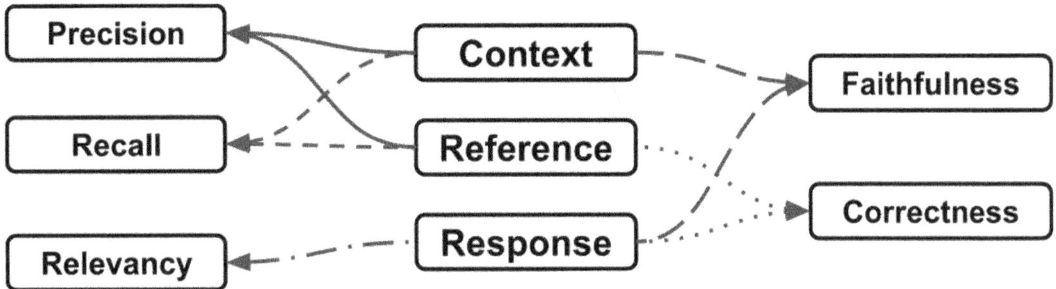

Figure 5-8. *The primary dependencies of the five explored RAGAS metrics on the context, reference, and RAG response. Context precision and recall depend on the context and reference, answer relevancy on the response, faithfulness on the response and context, and answer correctness on the response and reference. Most of the metrics depend on the user's prompt*

Without a ground-truth reference, these metrics cannot be evaluated because they rely on comparing the generated response and retrieved contexts against the correct answer.

This is the content of the Pandas DataFrame. Note how the answer correctness is high (0.8956) because the LLM's generated answer (*The capital of Egypt is Cairo*) to the user's question is correct.

```
user_input: What is the capital of Egypt?
retrieved_contexts: ['Cairo is the capital of Egypt.',
                  "Egypt's capital is very crowded."]
response: The capital of Egypt is Cairo.
reference: Cairo
context_precision: 0.9999999999
context_recall: 1.0
answer_relevancy: 0.9999999999999997
faithfulness: 1.0
answer_correctness: 0.8956099607502929
```

Correctness

Consider an error case in which the generated answer incorrectly states that New York is the capital of Egypt.

```
data = {"user_input": ["What is the capital of Egypt?"],
```

```
        "retrieved_contexts": [["Cairo is the capital of Egypt.",
                               "Egypt's capital is very crowded."]],
        "response": ["The capital of Egypt is New York."],
        "reference": ["Cairo"]}
```

In the resulting DataFrame, the answer correctness score is very low (e.g., 0.12). This is expected, as the answer correctness is computed by an LLM that compares the generated answer to the ground-truth reference (*Cairo*) and penalizes contradictions or factual mismatches.

```
user_input: What is the capital of Egypt?
retrieved_contexts: ['Cairo is the capital of Egypt.',
                    "Egypt's capital is very crowded."]
response: The capital of Egypt is New York.
reference: Cairo
context_precision: 0.9999999999
context_recall: 1.0
answer_relevancy: 0.9999999999999997
faithfulness: 0.0
answer_correctness: 0.12127478335971043
```

Faithfulness

Notice that the faithfulness score dropped from *1.0* to *0.0* because the generated answer does not align with the retrieved context. While the context states that *Cairo* is the capital of Egypt, the RAG response incorrectly asserts that it is *New York*. Since the answer is not supported by the retrieved evidence, faithfulness drops to zero.

If, however, the retrieved context also indicated that *New York* is the capital of Egypt, the answer would then be consistent with the context. In that case, the faithfulness score would rise to *1.0*.

```
data = {"user_input": ["What is the capital of Egypt?"],
       "retrieved_contexts": [["New York is the capital of Egypt.",
                               "Egypt's capital is very crowded."]],
       "response": ["The capital of Egypt is New York."],
       "reference": ["Cairo"]}
```

Once this change is applied, the faithfulness score rises to *1.0* because the generated answer is fully supported by the retrieved context. However, both context precision and context recall drop from *1.0* to *0.0*. This happens because these metrics are computed with respect to the reference answer.

```
user_input: What is the capital of Egypt?
retrieved_contexts: ['New York is the capital of Egypt.',
                     "Egypt's capital is very crowded."]
response: The capital of Egypt is New York.
reference: Cairo
context_precision: 0.0
context_recall: 0.0
answer_relevancy: 0.9999999999999997
faithfulness: 1.0
answer_correctness: 0.12128169742744838
```

Context Recall

Since the reference answer contradicts the RAG model's response and the retrieved context, the evaluation determines that no relevant context was retrieved, leading to very low scores for precision and recall. As in Figure 5-8, the context recall depends on the reference answer and the context. Let's change the reference answer to be New York:

```
data = {"user_input": ["What is the capital of Egypt?"],
        "retrieved_contexts": [["New York is the capital of Egypt.",
                                "Egypt's capital is very crowded."]],
        "response": ["The capital of Egypt is New York."],
        "reference": ["New York"]}
```

Since the reference answer matches the RAG model's output, the context recall score is *1.0*. However, the context precision score remains *0.0*. Why does this happen?

```
user_input: What is the capital of Egypt?
retrieved_contexts: ['New York is the capital of Egypt.',
                     "Egypt's capital is very crowded."]
response: The capital of Egypt is New York.
reference: New York
context_precision: 0.0
```

```
context_recall: 1.0
answer_relevancy: 0.9999999999999997
faithfulness: 1.0
answer_correctness: 0.8607662180985043
```

Context Precision

Recall emphasizes quantity, while precision emphasizes quality. In other words, recall is concerned with retrieving as many relevant items as possible, even if some are only loosely related. But precision is concerned with ensuring that items classified as relevant are highly accurate.

In the context of RAG evaluation, recall is analogous to retrieving any chunk that overlaps with the reference answer, even if the overlap is weak. Precision, on the other hand, requires that the retrieved chunks be strongly relevant to the reference.

In our case, the reference answer is simply *New York*, which shares only a fragment with the first retrieved chunk (*New York is the capital of Egypt.*). Because the similarity is weak, precision remains low even though recall is high.

Let's change the reference answer to be identical to the first retrieved chunk (*"New York is the capital of Egypt."*).

```
data = {"user_input": ["What is the capital of Egypt?"],
        "retrieved_contexts": [["New York is the capital of Egypt.",
                                "Egypt's capital is very crowded."]],
        "response": ["The capital of Egypt is New York."],
        "reference": ["New York is the capital of Egypt"]}
```

The semantic similarity between the reference and that chunk is now maximized. In this case, the chunk is judged fully relevant, and the context precision score becomes *1.0*.

```
user_input: What is the capital of Egypt?
retrieved_contexts: ['New York is the capital of Egypt.', "Egypt's capital
is very crowded."]
response: The capital of Egypt is New York.
reference: New York is the capital of Egypt
context_precision: 0.9999999999
```

```
context_recall: 1.0
answer_relevancy: 0.9999999999999997
faithfulness: 1.0
answer_correctness: 0.9583214550135921
```

Relevancy

Across all the previous variations, the answer relevancy metric remained at *1.0*. As shown in Figure 5-8, this metric depends on the RAG model's response and, for sure, the prompt. In earlier experiments, both remained focused on the same theme related to capitals and countries. Since they have a high semantic similarity, the relevancy score is perfect.

To illustrate how answer relevancy can be reduced, consider changing the generated RAG response to an unrelated topic, such as football.

```
data = {"user_input": ["What is the capital of Egypt?"],
        "retrieved_contexts": [["Cairo is the capital of Egypt.",
                                "Egypt's capital is very crowded."]],
        "response": ["Playing football."],
        "reference": ["Cairo"]}
```

Since this response is semantically far from the prompt, the answer relevancy score drops significantly to *0.125*.

```
user_input: What is the capital of Egypt?
retrieved_contexts: ['Cairo is the capital of Egypt.', "Egypt's capital is
very crowded."]
response: Playing football.
reference: Cairo
context_precision: 0.9999999999
context_recall: 1.0
answer_relevancy: 0.12536980173879675
faithfulness: 0.0
answer_correctness: 0.02083739465266503
```

5.7 Advanced RAG Architectures and Variations

So far, we have explored the simplest form of the RAG model architecture, called VectorRAG. When a user submits a query, the system converts it into a vector, searches a pre-indexed database for the most mathematically similar text chunks, and feeds those chunks directly into the LLM.

Beyond the foundational architecture, other specialized RAG models have been developed to enhance the retrieval and augmentation process. They aim to increase the factual quality and reasoning capabilities of the final response by adding more layers of intelligence. We will just have an idea about some of these new models.

5.7.1 GraphRAG

Moving beyond simple text chunks, another variation is GraphRAG. It organizes information into a knowledge graph of interconnected entities and relationships. Rather than just finding similar sentences, it understands how concepts are linked. Standard VectorRAG might fail if a specific relationship isn't explicitly stated in a single text chunk but is hidden in the connections between different pieces of information. GraphRAG excels at multi-hop reasoning.

This gives GraphRAG the ability to associate different but related entities. For example, how a specific cooking technique affects a chemical property in food, or how the heat from one car part can damage another. This allows the model to provide answers based on the big picture rather than just the most similar-sounding paragraph.

5.7.2 HybridRAG

Since VectorRAG and GraphRAG each have their own strengths and limitations, HybridRAG has emerged as a powerful variation that combines multiple retrieval strategies. In this architecture, VectorRAG is typically used to quickly identify the most semantically similar chunks within a massive dataset. Once those core pieces are found, GraphRAG takes over to explore the underlying relationships among those chunks, uncovering nuanced information that a simple similarity search would have missed.

If VectorRAG is the tool that finds you the right slice of pizza, and GraphRAG is the map that shows you how all the ingredients were sourced, HybridRAG is like having a gourmet chef who finds the best slice and then explains exactly how the flavors of the crust, sauce, and cheese work together to create the final meal.

This is how each RAG type works:

- **VectorRAG**: A waiter who finds the slice that tastes most like what you described, based on vibes and similarity.

- **GraphRAG**: A food critic who knows that this pizza, that chef, that region, that flour supplier, and can answer questions by traversing those connections.

- **HybridRAG**: Both the waiter and the critic working together to get the closest-matching slice and the full relational context of why it's the right one.

5.7.3 SelfRAG

SelfRAG (Self-Reflective RAG) is designed specifically to combat hallucinations. Unlike standard models that blindly process any retrieved slice of data, SelfRAG is trained to critique its own workflow.

Before generating a final response, the model evaluates the retrieved chunks for relevance, checks if its proposed answer is actually supported by the evidence, and assesses the overall utility of the output. If the retrieved information is deemed irrelevant or contradictory, the model can proactively trigger a new search or reject the data entirely.

Standard RAG uses in-context learning over retrieved documents with a fixed pipeline, whereas SelfRAG introduces self-evaluation and reflection mechanisms that allow the model to iteratively assess and improve both retrieval and generation.

This self-correcting loop ensures that the final response is not just a guess based on similar-sounding text but a factually grounded answer that has passed its own rigorous internal audit.

5.7.4 AgenticRAG

AgenticRAG is the most modern iteration of the architecture, where the system is no longer a passive follower of instructions but an active participant. In this model, the LLM acts as an agent equipped with a diverse set of tools and data sources. Instead of being limited to a single vector database, the agent can dynamically choose the best source for the task at hand, including performing a web search, executing code scripts for calculations, or querying external structured databases.

If the first set of retrieved chunks doesn't provide a complete answer, the agent recognizes the gap and autonomously decides to refine its strategy, perhaps by performing a second or third search using different terms or a different tool altogether. It effectively thinks before it speaks, verifying its own work and iterating until it has synthesized a high-confidence solution.

While AgenticRAG offers unmatched reasoning power, it introduces significant trade-offs in latency, cost, and reliability. Because the model must think and iterate through multiple reasoning steps, a single query can take several seconds to complete, making it much slower than a standard search.

Each of the steps performed by AgenticRAG may require additional LLM calls, which can rapidly increase API costs compared to a single-shot retrieval. Furthermore, without strict constraints, an agent can occasionally fall into infinite loops or suffer from agent drift, where it chooses the wrong tool for a task or becomes stuck searching for a perfect answer that may not exist.

CHAPTER 6

Agentic AI

The early usage of LLM-powered applications or chatbots was limited to receiving prompts from the user and generating text-based responses. In this setup, the user interacted with the application purely as an assistant, without the system having the ability to take independent actions. As LLMs often lacked the necessary knowledge to fully answer certain queries, developers began augmenting them with retrieved information from external sources through RAG.

Today, LLMs can be integrated into a new class of AI systems known as agentic AI, which go beyond passive assistance. These agents can plan, make decisions, and take autonomous actions by interacting with external tools, APIs, and environments to accomplish complex tasks on behalf of the user. This chapter introduces agentic AI, a subfield of artificial intelligence focused on using generative AI tools to build agents capable of performing autonomous tasks.

With the spread of the agentic AI systems, it became a challenge to access the other agents or tools required by the agent to accomplish a task. To address this challenge, we will explore two widely accepted protocols designed to unify communication. The first protocol is Agent2Agent by Google, which is responsible for communication among the agents. The second protocol is the Model Context Protocol, developed by Anthropic, responsible for the communication between an agent and external tools and resources.

6.1 Introduction

LLMs have proved their remarkable capabilities in understanding user prompts, even those containing thousands of tokens, and generating relevant, coherent responses. In typical usage, a human interacts with the LLM, provides a prompt, and receives a one-time response. This interaction model is well-suited for tasks that are clear to the

© Ahmed Fawzy Gad 2026
A. F. Gad, *Transformers and Large Language Models*, https://doi.org/10.1007/979-8-8688-2785-3_6

model and can be completed in a single step, such as fixing a bug in a code snippet or generating an image of a blue sky. In such cases, the user is responsible for crafting the prompt and selecting the appropriate model to achieve the desired output.

However, single-shot responses are not always sufficient. This is because the model may not have enough resources or knowledge to handle the user's request. Moreover, the goal in many real-world scenarios is not just to give the user an answer but to act, which is often through a sequence of interdependent steps or small actions. For instance, hotel reservations may require reasoning, planning, tool use, or taking multiple decisions. Since the language model is intelligent in analyzing the natural language but does not have the ability to take actions, it uses an agency, which is the agentic AI, to become more practical and useful.

Traditionally, solving a task requires manual planning, step-by-step reasoning, and coding. For example, consider the task of booking a bus to get to work. This typically involves a sequence of decisions and actions, such as:

1.   Identifying your current location

2.   Determining the location of your workplace

3.   Finding the best bus route.

4.   Finding the nearest bus station

5.   Booking a ticket through the appropriate service

For each new task, a similar reasoning process must be repeated, often requiring explicit programming or user intervention. In conventional systems, this logic is typically hardcoded, meaning the system follows predefined steps set by the developer.

Agentic AI represents a paradigm shift from this manual, static approach. These systems are dynamic and capable of performing autonomous reasoning, as in Figure 6-1. Instead of relying on hardcoded instructions, an agentic AI system can automatically infer the necessary steps to complete a task. Given a high-level goal (e.g., "book a bus to work"), it can identify subtasks, gather relevant information, and execute the required actions without explicit guidance for each step. This autonomy significantly reduces the need for manual planning and allows systems to adapt to new contexts or objectives more flexibly.

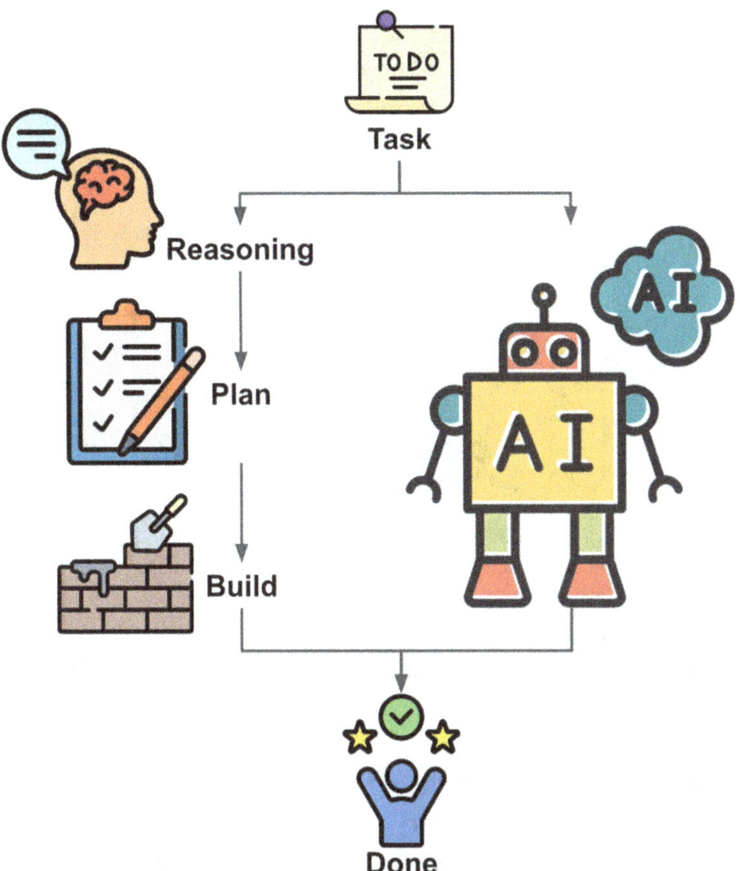

Figure 6-1. *Agentic AI fulfills tasks autonomously without requiring manual planning. Given only a high-level task description, it can infer and execute the necessary steps to complete the task*

Traditionally, people often rely on agencies to handle tasks that require specific knowledge or effort, such as booking a cheap flight. Finding the best deals usually involves navigating various websites and understanding pricing strategies. An agency simplifies this by offering not only expertise but also flexible payment options and more. If you're unfamiliar with how to book a ticket yourself, you'd likely turn to such a service offered by the agency to handle it for you. Instead of doing all the work personally, people prefer to delegate these tasks to professionals they trust. They don't necessarily want to know all the details as long as the job gets done right.

Agentic AI works in a similar way, as illustrated in Figure 6-2. For example, you might use an LLM to classify emails into spam or ham. In this case, the LLM acts like an agency: you delegate the task and trust it to perform accurately, without needing to understand the full technical process behind it.

Figure 6-2. *Just as people rely on travel agencies to handle the complexity of booking flights, agentic AI uses LLMs to perform tasks on our behalf without managing the technical details such as email classification*

Agentic AI systems are designed to take autonomous actions to accomplish multi-step tasks. These systems consist of agents, each responsible for performing a specific subtask. Like functions or methods in a well-architected program, each agent operates based on a clearly defined specification of expected inputs and outputs. Importantly, agents function autonomously, without requiring constant human supervision or intervention.

Much like microservices in distributed systems, agentic AI leverages an orchestrator, which is a central component responsible for achieving the high-level goals through subtasks. This is by thinking and planning how the task could be accomplished through a series of subtasks, then finding the appropriate agents for each subtask and ensuring that the overall objective is achieved at the end. Usually, the agents think of using an external LLM model. It is not necessarily an LLM, but any language model capable of giving the right reasoning.

For the orchestrator to assign tasks for the agents, such agents must have a clear description of their capabilities. If the orchestrator has a task, then it uses the capabilities of the known agents to decide which agent is the best for this task.

For example, consider a user interested in going kayaking next week. As in Figure 6-3, an agentic AI system could autonomously:

- Monitor the weather forecast,

- Identify the best time slots on sunny days,

- Explore available kayak rental options,

- And initiate a reservation pending user confirmation.

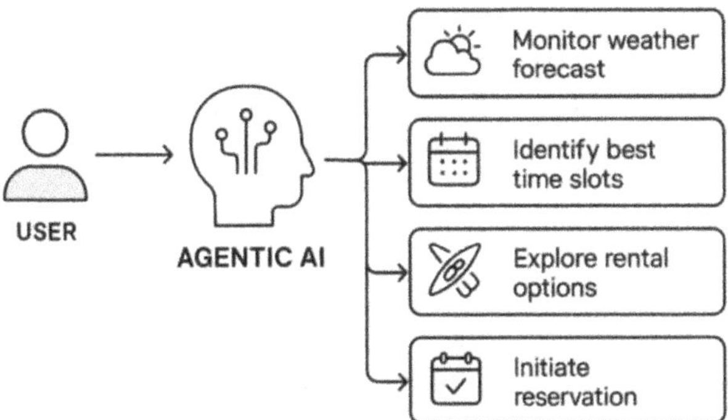

Figure 6-3. *Agentic AI coordinates multiple specialized agents to autonomously complete tasks on behalf of the user*

This is why agentic AI goes far beyond simple question answering. A group of agents work together to achieve the user's goal and not just give text responses. The above example uses a digital environment, but in real life, it might be extended to physical agents.

Autonomous decision-making is a core feature of agentic AI, but it is not the only defining characteristic. Agentic systems often have access to specialized resources and services that enhance their ability to perform complex tasks. Each agent is typically equipped with domain-specific tools or data sources. For example, one agent may have access to a list of local kayak rental shops, while another may connect to real-time weather data.

In contrast, traditional LLMs rely primarily on their pretrained knowledge and, in some cases, limited access to public resources. Even when integrated with external data, LLMs are generally constrained to generating responses, leaving the user responsible for taking action that happens after a lengthy back-and-forth conversation.

Agentic AI makes taking actions simpler. While each agent acts independently and uses its own resources, the underlying language model remains central to the system. The LLM acts as the reasoning engine, responsible for interpreting instructions, making decisions, and selecting appropriate tools or data sources. In essence, the LLM "thinks" while the agent "acts" using the resources available to it. An agent might not necessarily have advanced AI embedded into it, but it knows how to use the AI tools to accomplish the user's tasks.

> *An agent may have access not only to an LLM but also to a broader set of resources, including MCP servers, other agents, retrieval-augmented generation (RAG) systems, and various external tools or services. This expanded access allows the agent to perform more complex reasoning, fetch relevant knowledge, and coordinate with specialized components to fulfill diverse user requests.*

This architecture is conceptually similar to RAG, where an LLM retrieves relevant documents or data before generating a response. In agentic AI, however, this idea is extended further as the agents not only retrieve information but also execute actions based on that information to fulfill complex goals.

Creating an agent is not necessarily complex. A simple agent can be created by prompting a large language model (LLM) to respond to queries in a structured format. For example, the prompt could be:

> *I will ask you a question, and your job is to respond with either "Yes," "No," or "Maybe."*

If the input is *"Is it sunny today?"*, the model will return a constrained response. Because the LLM is performing a specific task in a repeatable and structured way, it can be considered a basic agent even if it does not access any external tools or knowledge beyond the LLM's built-in capabilities.

In the earlier discussion, LLM applications were expected to have direct access to all the external tools and agents they needed to perform tasks. Achieving this typically required building custom APIs for each individual service such as email, weather data, or databases. This approach can be time-consuming and error-prone, especially as the number of tools grows.

As agentic AI systems grow more complex, there is a growing need for standardized ways to enable communication between agents, tools, and external resources. To address this, open protocols have been introduced to provide unified communication mechanisms. Two of the most common protocols are:

1. **Agent2Agent (A2A) Protocol**: Defines a standardized method for communication between autonomous agents.

2. **Model Context Protocol (MCP)**: Provides a unified interface for agents, typically LLMs, to interact with tools and external resources.

Anthropic AI introduced the model context protocol (MCP) as an elegant and open-standard solution for enabling language models to interact with external services. MCP allows an LLM to access various tools and resources in a structured, consistent way without requiring a custom API for each interaction. This significantly reduces development overhead and promotes interoperability across different systems and ecosystems.

MCP is designed around a single-agent model, focusing on structured communication to allow the LLM to access external tools or resources. In contrast, to support communication between multiple autonomous agents with distinct capabilities, Google proposed the Agent2Agent (A2A) protocol. A2A facilitates coordination among agents working collaboratively on complex, multi-step tasks.

While MCP and A2A serve different purposes, they are highly complementary, as in Figure 6-4. MCP standardizes tools and resources access by a single agent, whereas A2A enables collaboration between multiple agents. The user interacts with an agent that maintains a directory of other agents and their capabilities. Based on the user's prompt, this agent may delegate the task to another, more specialized agent using A2A communication. To generate an accurate response, the helper agent may require external information or tools it has access to (e.g., current weather data). It retrieves it by invoking the appropriate external tool or resource via the MCP.

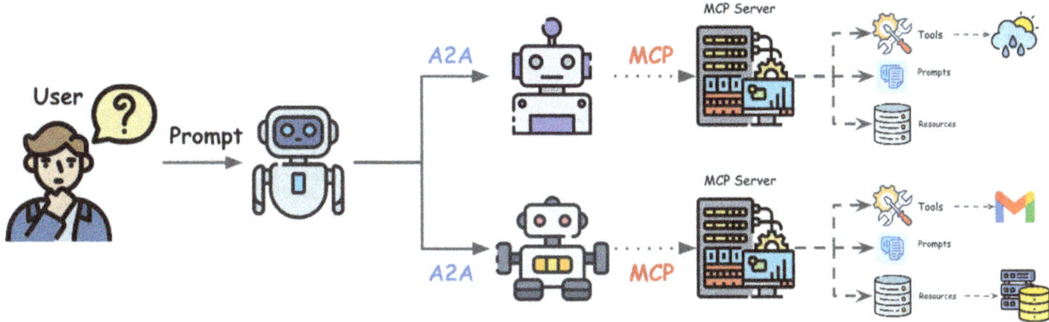

Figure 6-4. *Agent2Agent (A2A) and model context protocol (MCP) work in a complementary manner. A2A enables communication and coordination between agents, while MCP allows an agent to access external tools or resources to expand its contextual understanding and capabilities*

In the following two sections, we will explore the MCP and the A2A protocols through practical examples. These examples will utilize the official Python SDKs provided by Anthropic and Google, demonstrating how these protocols enable tool-augmented and collaborative AI agents. Since the MCP was introduced prior to the A2A protocol, we will begin our discussion with MCP.

6.2 Model Context Protocol (MCP)

Many AI models provide a range of services, but each often comes with its own interface and a dedicated API, making integration with external tools and systems complex. Typically, each model provider defines a unique interface for accessing their models, requiring developers to tailor their integration efforts for each case.

To address this challenge, Anthropic introduced the model context protocol (MCP) in November 2024, as outlined in their blog post *Introducing the Model Context Protocol*. As the name implies, the MCP supplies additional context to models, such as large language models (LLMs), enabling them to access more relevant information and thus generate more accurate and informed responses.

MCP is a universal, open-standard protocol designed using a client-server architecture to streamline communication between models and external tools or data sources, as illustrated in Figure 6-5. By offering a standardized interface, MCP greatly simplifies the process of building, connecting, and integrating tools across different models. Developers will also save time maintaining separate interfaces.

This is similar to how older computers require separate ports for the keyboard, mouse, and speakers. These individual interfaces were eventually replaced by the Universal Serial Bus (USB) port, which provides a unified connection standard.

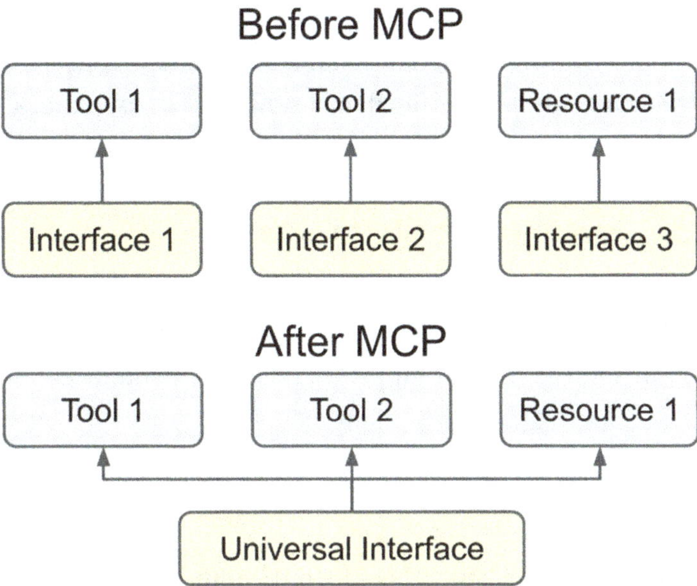

Figure 6-5. *The model context protocol (MCP) provides a universal interface for accessing any MCP-compliant tool or resource, eliminating the need for dedicated, tool-specific interfaces*

If you are developing your own data source or tool and wish to make it accessible to external systems, you can implement an MCP server to expose your resources. Conversely, to access tools or services, you can use an MCP client, provided those tools are hosted behind an MCP-compliant server.

An MCP server provides three primary capabilities to support interaction with external tools and resources in a standardized way:

1. **Resources**: These are data elements that can be fetched by MCP clients. They are typically served in file-like formats. Resources might include structured data (e.g., user profiles, logs, database records) or unstructured content (e.g., documents or articles) that the model can read and reason over.

2. **Tools**: Tools represent executable functions exposed by the MCP server to do actions. Each tool performs a specific task, such as retrieving weather information, accessing sports results, sending an email, or querying a database. Tools are defined with a name, a description, and a schema for their input and output parameters. These descriptions help AI models determine when and how to use each tool effectively.

3. **Prompts**: Prompts are pre-defined, customizable templates designed to assist users with common tasks. These templates can include input fields that are filled in at runtime. For example, a prompt template might be *"Return the weather temperature for the day {DATE}"*, where {DATE} is a placeholder that the user or the model can replace with a specific value.

AI applications such as Anthropic's Claude Desktop come equipped with an MCP client. This client is responsible for connecting to one or more MCP servers and accessing the tools, resources, and prompts they provide. To interact effectively, the MCP client must:

- Discover tool capabilities by retrieving tool descriptions and documentation.

- Select appropriate tools based on the task requirements.

- Handle data exchange, such as receiving resources from the server or submitting input to tools and prompts.

Another key advantage of using the MCP to access external resources, rather than relying on traditional APIs, is the reduced maintenance overhead. Traditional APIs require developers to be constantly aware of accepted parameters. If an API changes, such as removing or renaming a parameter, developers must manually update their code to avoid errors or failures.

In contrast, the MCP system handles such changes dynamically. Each MCP server describes its available tools along with the current set of accepted parameters. This information is communicated back to the AI model via the MCP client, which constructs requests based on the latest parameter definitions. If the server updates its tools by either adding, removing, or modifying parameters, it automatically communicates the updated specification. As a result, the system adapts in real time, eliminating the need for manual interface maintenance.

Consider a scenario where you have documents stored in a Google Drive folder and you want to make it usable by AI applications to draft an email summarizing their content. Traditionally, this would require writing custom code to integrate with both Google Drive and an email service like Gmail.

With the MCP, this process becomes significantly simpler. You create an MCP server to expose a standardized interface to access both Google Drive and the email service. The AI application or the LLM, equipped with an MCP client, can then connect to the MCP server to read the files and draft an email.

As a summary, the MCP ecosystem consists of three main components, as illustrated in Figure 6-6:

1. **MCP Host**: The front-end system through which the user interacts. This is typically powered by an LLM, such as Anthropic Claude. It represents the host where the MCP client runs.

2. **MCP Client**: The middleware responsible for connecting the AI application to the MCP server. It facilitates tool discovery, invocation, and data exchange.

3. **MCP Server**: The backend that hosts and exposes tools, resources, or prompt templates in a standardized format.

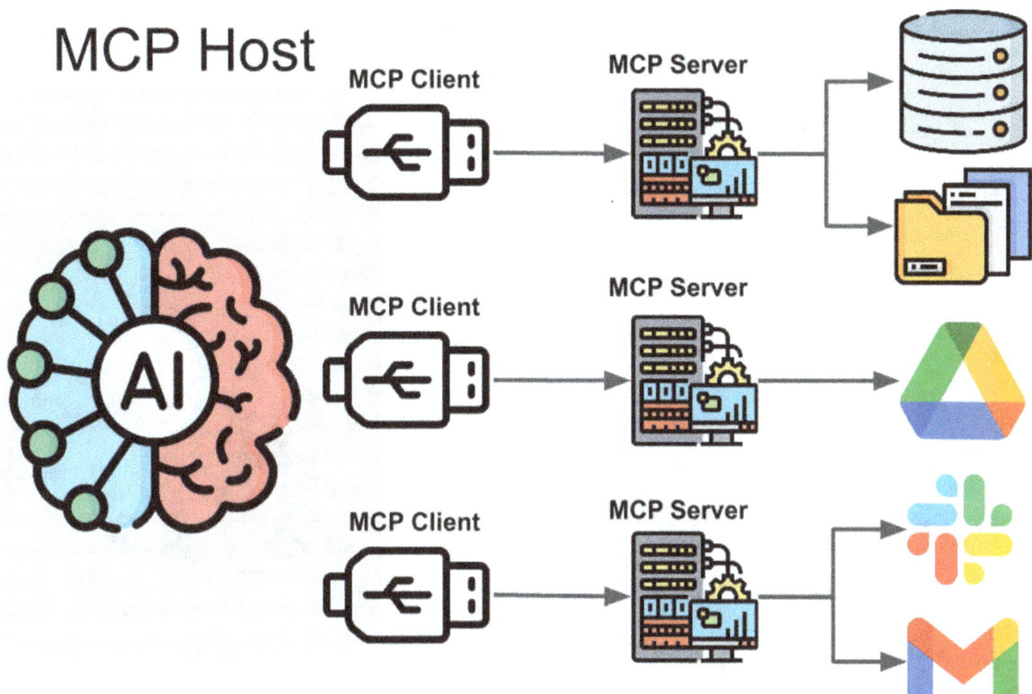

Figure 6-6. *The connection between the MCP host, MCP client, and MCP server. Each MCP server includes its own MCP client to manage interactions. The MCP host integrates AI tools, such as an LLM, which acts as the system's brain*

The detailed interaction workflow among these components is illustrated in Figure 6-7 and proceeds as follows:

1. The user enters a prompt into the AI application. A prompt template might be used.

2. The MCP client forwards the prompt to the LLM, which interprets it and determines whether external tool usage is needed.

3. If a tool is required, the LLM selects the appropriate tool from those available on the MCP server.

4. The MCP client then sends a request to the MCP server to execute the selected tool.

5. The MCP server processes the request and returns the result to the MCP client.

6. The MCP client delivers the tool's output back to the LLM.

7. The LLM incorporates the MCP server tool's output into the final response and send it back to the MCP client.

8. The MCP client shows the response back to the user.

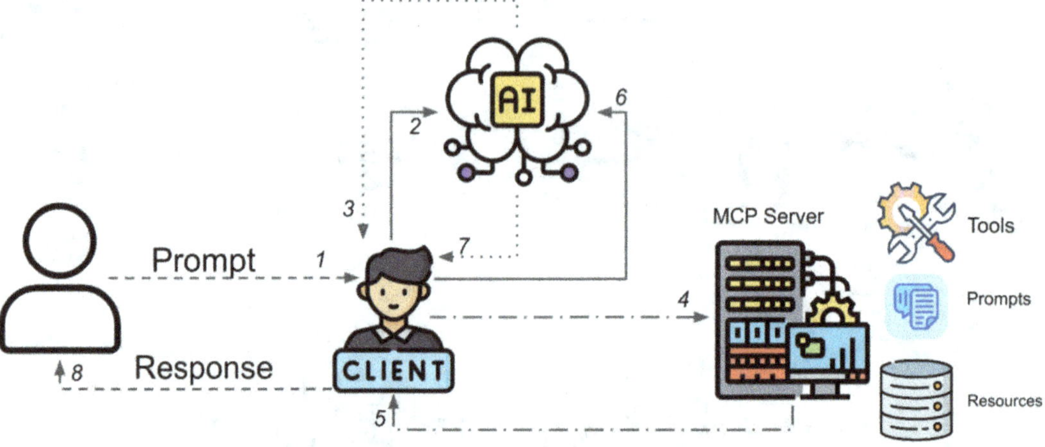

Figure 6-7. *Interaction workflow between the AI application, MCP client, and MCP server. The MCP server exposes tools, resources, and prompt templates, which the MCP client accesses on behalf of the AI application to fulfill user requests*

As long as an AI application includes an MCP client, it becomes significantly easier to integrate your tools and resources across different platforms. For example, suppose that Google Gemini, Anthropic Claude, and OpenAI ChatGPT all support MCP clients. In that case, you can simply create an MCP server that exposes your tools through the standardized MCP interface. Once available, these tools can be accessed by any of the supported applications without the need to write separate, custom APIs for each platform.

6.2.1 MCP Python SDK

Anthropic has released an implementation of the MCP along with SDKs available in multiple programming languages. In this section, we will explore how to use the MCP Python SDK to build an MCP server. The server will provide a tool that fetches the latest football match result for a club based on its name.

Since the Anthropic Claude Desktop application is developed by the creators of MCP, it is highly compatible with the protocol and serves as an excellent MCP host. Claude includes a built-in MCP client, allowing seamless connections to MCP servers.

The source code for the MCP Python SDK is available on GitHub: `https://github.com/modelcontextprotocol/python-sdk`

For building an MCP server, the following steps outline the workflow to follow:

1. Prepare the Python project using the uv tool.

2. Build the MCP server with the necessary tools and prompts.

3. Run the MCP server locally on your machine.

4. Install Claude Desktop, the LLM client compatible with MCP.

5. Configure Claude Desktop to connect to our MCP server.

6. Exploring the MCP server tools in Claude.

7. Interact with the MCP server tool by entering prompts in Claude.

Initialize the Python Project Using uv

We will use the uv Python project manager to manage our environment and dependencies. Unlike the traditional workflow where you manually create a virtual environment and ensure it is activated before use, uv simplifies this process by automatically handling virtual environment activation on the fly. This allows you to focus more on coding and less on setup.

To initialize a new project named sum using uv, execute these commands:

```
uv init sum
cd sum
```

Next, create a virtual environment:

```
uv venv
source .venv/bin/activate
```

Finally, add the necessary dependencies. Only the mcp library with its CLI extension is needed.

```
uv add "mcp[cli]"
```

The uv init command sets up the project directory and generates configuration files, including pyproject.toml. This file contains important project metadata such as dependencies, Python version requirements, and the project name.

Once the project setup is complete, create a new Python script named sum.py in the project directory. This script will be used to develop the MCP server. You are now ready to begin coding your MCP server.

Build a Toy MCP Server

To begin, we will build a simple MCP server that provides a single tool for adding two numbers. This toy example allows us to focus on understanding the structure and usage of the MCP SDK without being distracted by the complexity of a real-world task of getting the match results.

There are two approaches to building an MCP server:

1. Low-level approach using the mcp.server.lowlevel.server. Server class, which offers fine-grained control over server behavior.

2. High-level approach using the `mcp.server.fastmcp.FastMCP` class, which simplifies development through sensible defaults and streamlined configuration. The *FastMCP* framework is maintained through this GitHub repository `https://github.com/jlowin/fastmcp` and used by the official Anthropic MCP Python SDK. Its documentation

The FastMCP framework is actively maintained on GitHub (`https://github.com/jlowin/fastmcp`) and is integrated into the official Anthropic MCP Python SDK. Comprehensive documentation is available at `https://gofastmcp.com`.

For the purposes of this introduction, we will use the `FastMCP` class to quickly get started. The first step is to create an instance of this class.

```python
from mcp.server.fastmcp import FastMCP

mcp = FastMCP(name="Sum Numbers",
              instructions="You are a virtual assistant that helps users
              add two numbers. Use the provided tool to perform the
              addition and return the result.")
```

The `FastMCP` class accepts several optional parameters in its constructor, but none are required—at least at the time of writing this book. In our example, we use the following two parameters:

1. `name`: Specifies the name of the MCP server. This name can help identify the server, especially when multiple tools or services are available.

2. `instructions`: Provides high-level guidance to the language model. This acts as a system-level prompt that gives the model context about its role and how it should behave when interacting with users.

As explained earlier, an MCP server can expose resources, tools, and prompts. We will begin by defining a single tool named `add_two_numbers`, as shown in the upcoming code example.

Tools in MCP are defined using standard Python functions but must be annotated with the `@mcp.tool` decorator. In this context, `mcp` refers to the instance of the `FastMCP` class we previously created.

The add_two_numbers tool accepts two parameters:

1. first_num: The first number to add.

2. second_num: The second number to add.

The tool simply returns the sum of these two numbers.

```python
@mcp.tool()
def add_two_numbers(first_num: float, second_num: float) -> float:
    """Add two numbers.

    Args:
        first_num: The first number to add.
        second_num: The second number to add.

    Returns:
        The sum of the two numbers.
    """

    result = first_num + second_num

    return result
```

Important   Proper documentation is critical when defining tools. This is how the LLM becomes aware of the tool and understands how and when to use it. The docstring should clearly describe the purpose of the tool, the expected arguments, and any return values. A well-documented tool improves the model's ability to call it accurately in response to user prompts.

Run the MCP Server

Once the MCP server is created, the next step is to run it. This is done by calling the server's run() method within your Python script. The run() method accepts an optional argument called transport, which specifies the communication protocol used between the MCP server and the MCP client.

The available transport protocols are:

1. `stdio`: Communication over standard input/output streams. Use it if the MCP server and client run on the same machine and communicate directly via process pipes. This is usually used only for local development.

2. `sse`: Server-Sent Events protocol used for streaming updates.

3. `streamable-http`: An HTTP-based streaming protocol. Usually used in production where the server and client communicate interactively over a network.

Choosing the appropriate transport depends on your deployment environment and client capabilities. In our example, both the MCP server and client will run locally so we can simply use the `stdio` option.

```python
if __name__ == "__main__":
    mcp.run(transport='stdio')
```

At this point, the MCP server script is complete and ready to run. You have two main options to execute it. The first option is using an IDE where you can run the script directly. Just make sure the virtual environment associated with the project is properly activated to ensure all dependencies (such as `mcp`) are available.

The better option is to use the `uv` tool, which is a simpler and cleaner approach, to run the script. It implicitly activates the virtual environment for you. This eliminates the need to manage environment activation manually.

```
uv run python sum.py
```

Using `uv` not only streamlines project setup but also ensures consistency across development workflows by managing the environment automatically.

Install Claude Desktop

Once the MCP server is ready and functioning, the next step is to install the Claude Desktop application. Claude Desktop serves as the interface where you can interact with the LLM and connect it to your local MCP server.

Claude is developed by Anthropic, the creators of the MCP, and is fully compatible with the MCP SDK. It comes with a built-in MCP client, which allows it to automatically discover and interact with any running MCP server on your machine.

At the time of writing, Claude Desktop is available for macOS and Windows. You can download the latest version from `https://www.anthropic.com/claude` *or* `https://claude.ai/download`*.*

Configure Claude Desktop

To enable Claude Desktop to interact with the MCP server we created, we need to configure it by editing a file named `claude_desktop_config.json`. This file holds the configuration for Claude Desktop, including registered MCP servers.

You can find this JSON configuration file path by following these steps:

1. Go to Claude Desktop

2. Go to Settings.

3. Select the Developer option.

4. Click Edit Config. This will take you to the location of the JSON file.

On macOS and Linux, the file is usually located at:

```
~/Library/Application\ Support/Claude/claude_desktop_config.json
```

On Windows, the file can be found at:

```
%APPDATA%\Claude\claude_desktop_config.json
```

Open the file with any text editor of your choice, and insert the following configuration snippet to register your MCP server:

```
{
  "mcpServers": {
    "sum": {
      "command": "/Users/ahmedgad/.local/bin/uv",
      "args": [
        "--directory",
        "/Users/ahmedgad/Desktop/LLMs/mcptest/sum",
        "run",
        "sum.py"
      ]
    }
  }
}
```

Make sure to replace the directories with the actual path to your uv executable and project directory where the sum.py script exists.

Once the configuration file is updated, restart the Claude Desktop application to apply the changes and enable the newly registered MCP server.

Explore the MCP Server in Claude

In the main interface of the Claude Desktop application, you will find a tools icon below the prompt textbox as in Figure 6-8. Click on it.

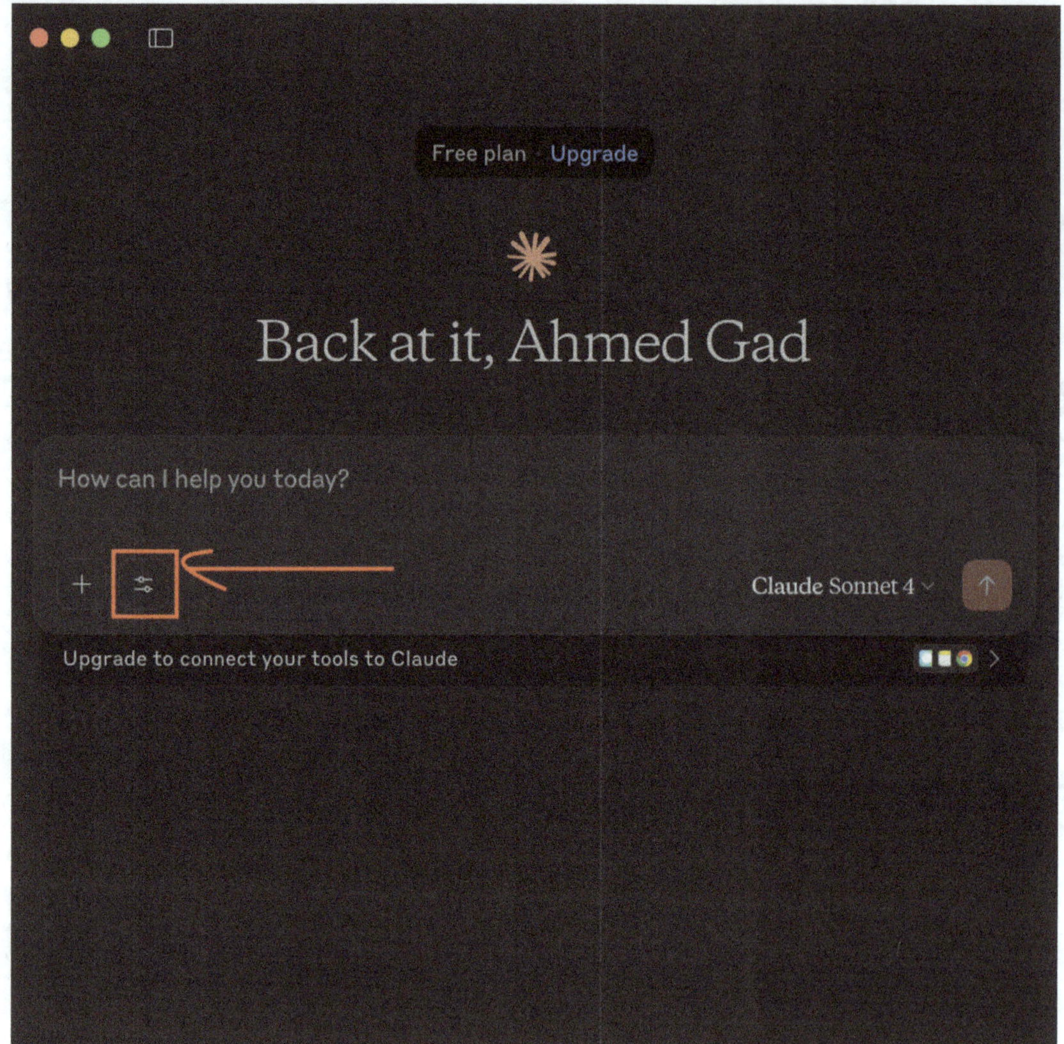

Figure 6-8. *The Claude Desktop application main interface. The Tools icon provides access to configured MCP servers*

After clicking the Tools icon, a popup menu appears showing several options, as in Figure 6-9, including MCP servers detected by the MCP client. In this example, there is a single MCP server named *sum*.

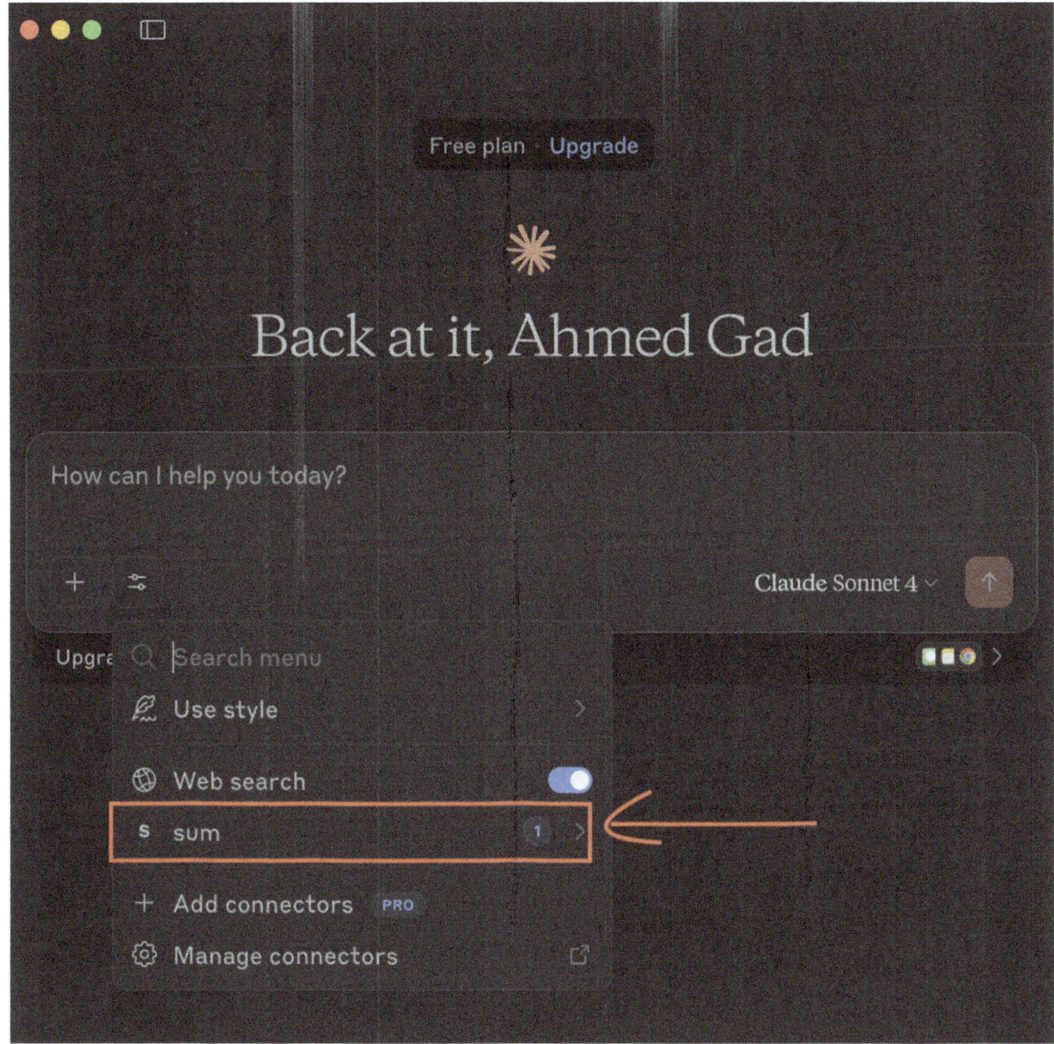

Figure 6-9. *Clicking on the Tools icon displays a menu of available tools in Claude, including the registered MCP servers*

Clicking the *sum* MCP server opens a new child window displaying the list of tools supported by the server. As shown in Figure 6-10, there is only one tool available, named `add_two_numbers`.

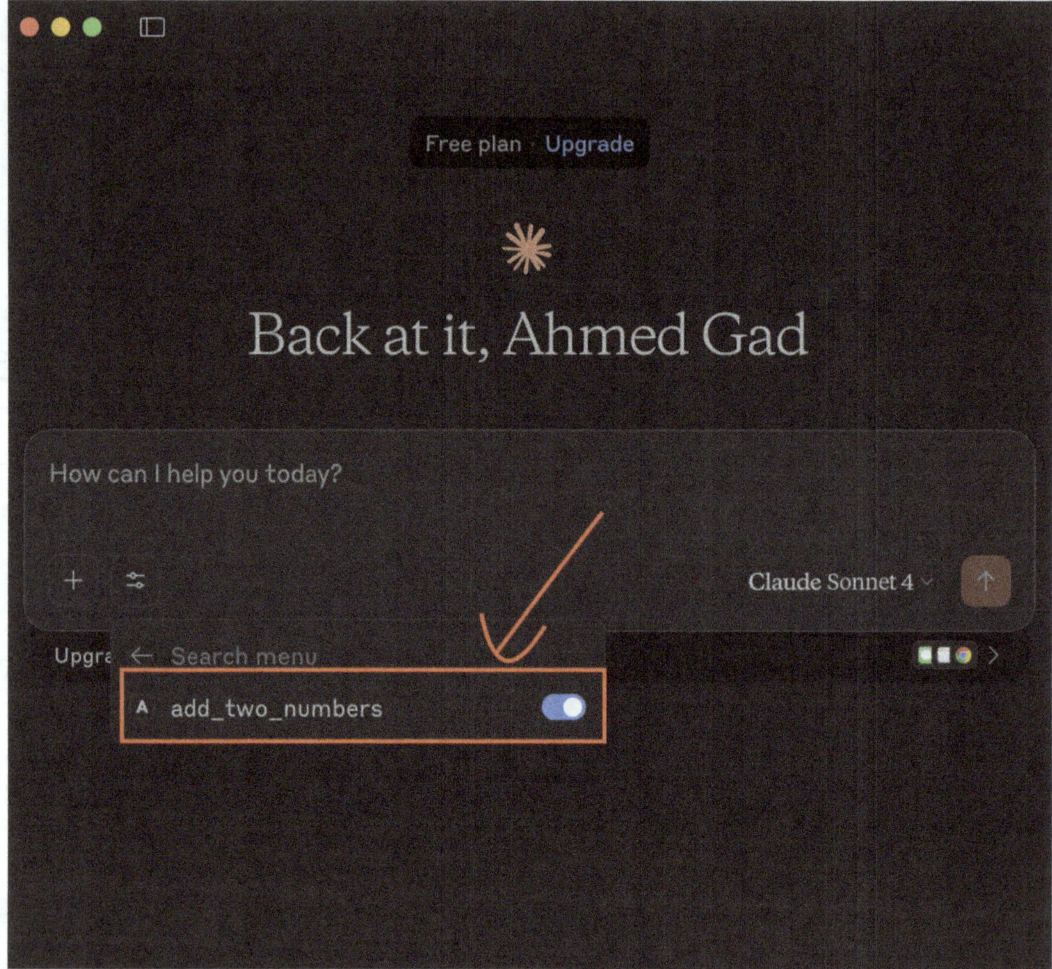

Figure 6-10. *The list of tools supported by the MCP server*

At this stage, we have confirmed that the MCP server is running and successfully detected by the MCP client within Claude Desktop. We have also verified that all tools exposed by the MCP server are visible in Claude's interface. The next step is to interact with Claude using prompts that will trigger the tool inside the MCP server.

Interact with the MCP Server

We are now ready to use Claude with the external tools provided by our MCP server. The only way to trigger a tool inside the MCP server is through user prompts that are relevant to the functionality of those tools.

Thanks to the LLM's powerful natural language understanding, it can automatically determine whether a given prompt can be addressed by one of the connected MCP server's tools. This decision is based on the tool's docstring, which describes its purpose and parameters. If an appropriate tool is identified, the LLM will invoke it to generate a response. This highlights the importance of writing clear, descriptive docstrings for all MCP server tools.

> *Remember, the primary purpose of our MCP server is to extend the LLM's capabilities by providing access to a variety of external tools and resources to enrich and extend its context beyond its trained knowledge. This supplies the model with more current and detailed information, enabling it to generate more accurate and relevant responses.*

Since the only tool in the MCP server is about summing two numbers, let's use a relevant prompt such as:

```
What is the result of summing 7 and 4?
```

When a prompt is entered into Claude, the following steps occur:

1. The MCP client discovers the available MCP servers (i.e., tools and resources) configured in Claude.

2. Claude analyzes the registered tools to determine if any are relevant to the prompt.

3. Once Claude identifies an appropriate tool, it requests permission from the user to invoke the selected MCP server tool (see Figure 6-11). The user can grant permission by clicking *Allow once* or *Allow always*.

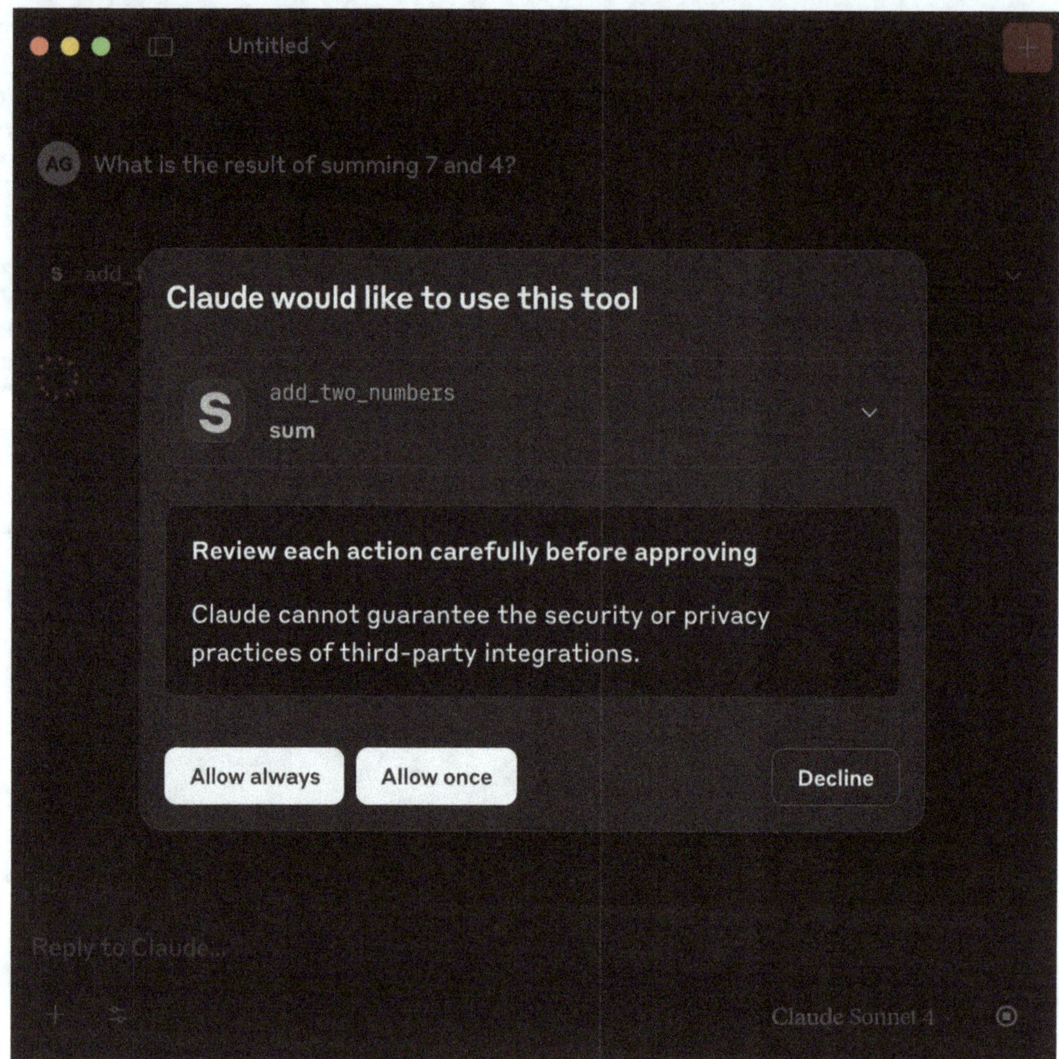

Figure 6-11. *Claude requests user permission to invoke a tool from the local MCP server*

4. After permission is granted, Claude (the LLM) drafts a structured request for the selected MCP server tool. Since the tool accepts 2 arguments called `first_num` and `second_num`, this is the JSON request.

```
{
   `first_num`: 7,
   `second_num`: 4
}
```

5. This request is forwarded to the MCP server via the MCP client.

6. The MCP server executes the tool and generates the output.

7. The output is sent back to the MCP client.

8. The MCP client forwards the result to Claude.

9. Claude incorporates it as context to generate the final response delivered to the user.

The response returned from Claude, shown in Figure 6-12, displays the sum of the two numbers as *11*. When Claude utilizes an MCP tool, the name of the selected tool appears below the prompt. This name can be clicked to expand a detailed view, revealing the request sent to the tool and the response received.

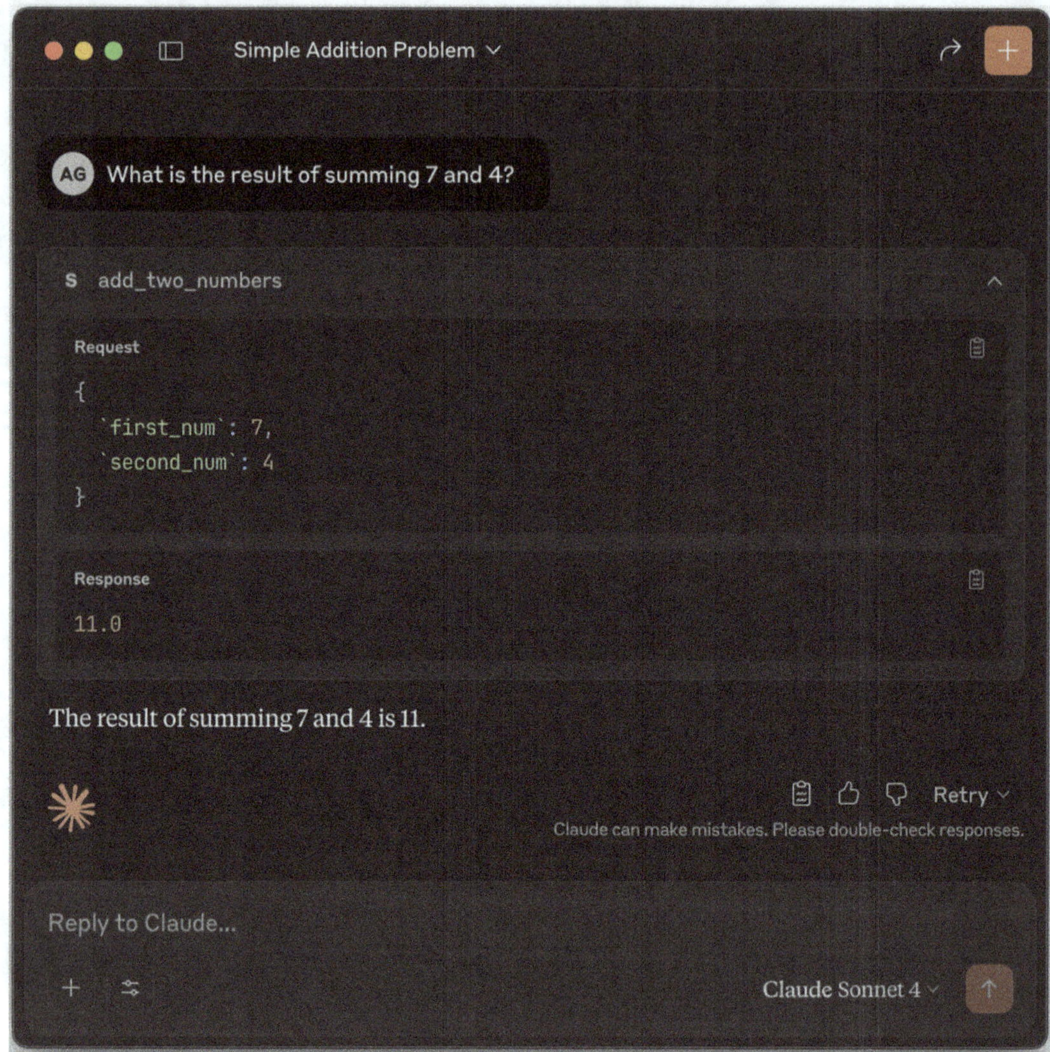

Figure 6-12. *Claude generates a response based on the context provided by the invoked MCP server tool. It also allows the user to view the tool request and response details*

Before completing this toy example, let's extend the server to add a prompt.

Add a Prompt to the MCP Server

To define a prompt in the MCP server, simply create a standard Python function and decorate it with @mcp.prompt(), as shown in the code below. A prompt acts as a reusable template designed to save users time by avoiding repetitive manual entry, especially for common tasks. It allows the user to provide only the necessary inputs.

```python
@mcp.prompt()
def request_two_numbers(user_name: str, first_num: float, second_num:
float) -> str:
    """Just a quick prompt to the LLM model to help the user use our tool
    easily"""
    return f"You are an assistant for the user {user_name}. Given the
    accepted two numbers {first_num} and {second_num}, what is their
    summation? Please greet the user by its name."
```

The prompt, request_two_numbers, takes three arguments:

1.   user_name: The name of the user.

2.   first_num: The first number.

3.   second_num: The second number.

These inputs are used to dynamically construct a prompt that is returned to Claude for processing.

You need to update the sum.py script with the new prompt's code above. Both the MCP server and Claude must be restarted for the change to take effect. Then the prompt will be recognized by Claude as in Figure 6-13.

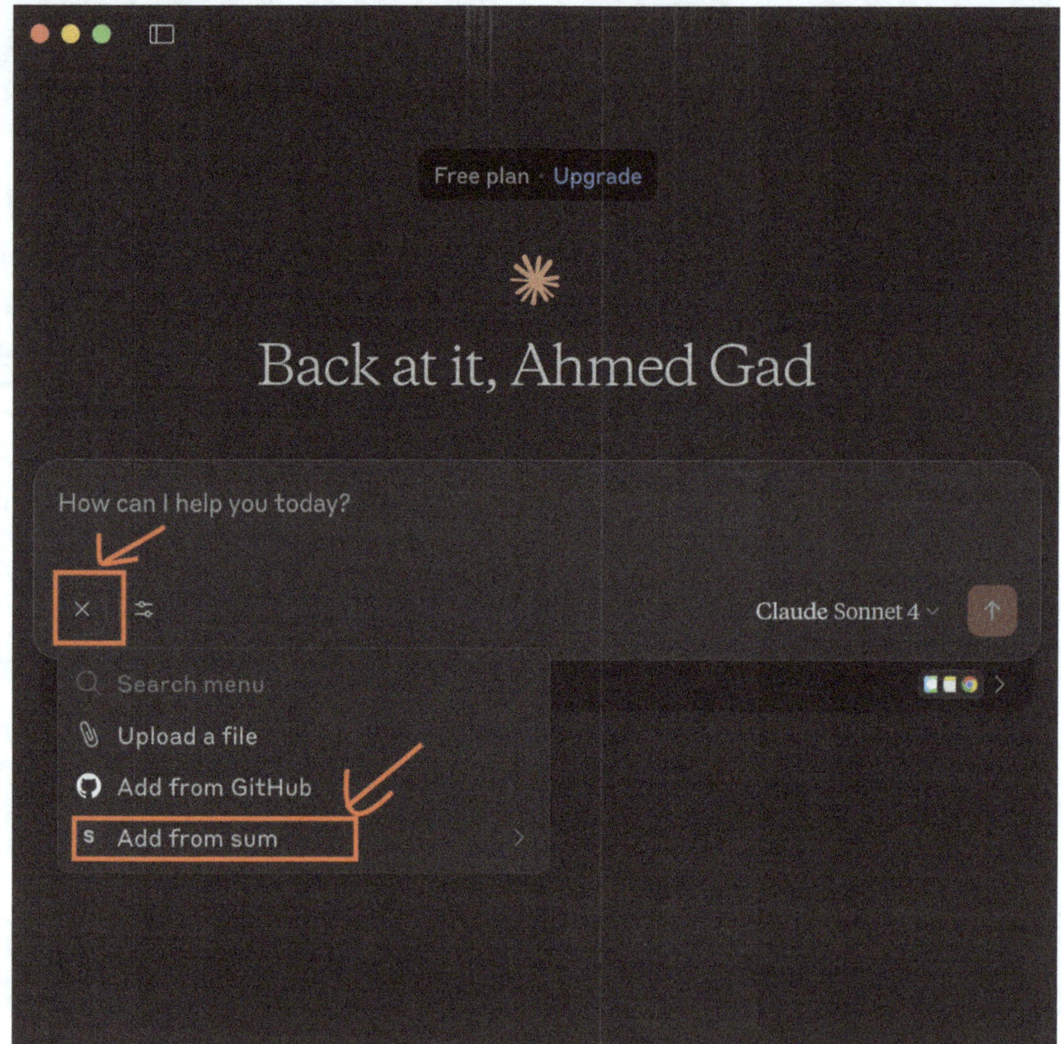

Figure 6-13. *Clicking on the plus icon next to the tool's icon opens a menu with options to access the prompts provided by the connected MCP servers*

By clicking on the option corresponding to the *sum* MCP server, a list appears with all available prompts it supports, as shown in Figure 6-14.

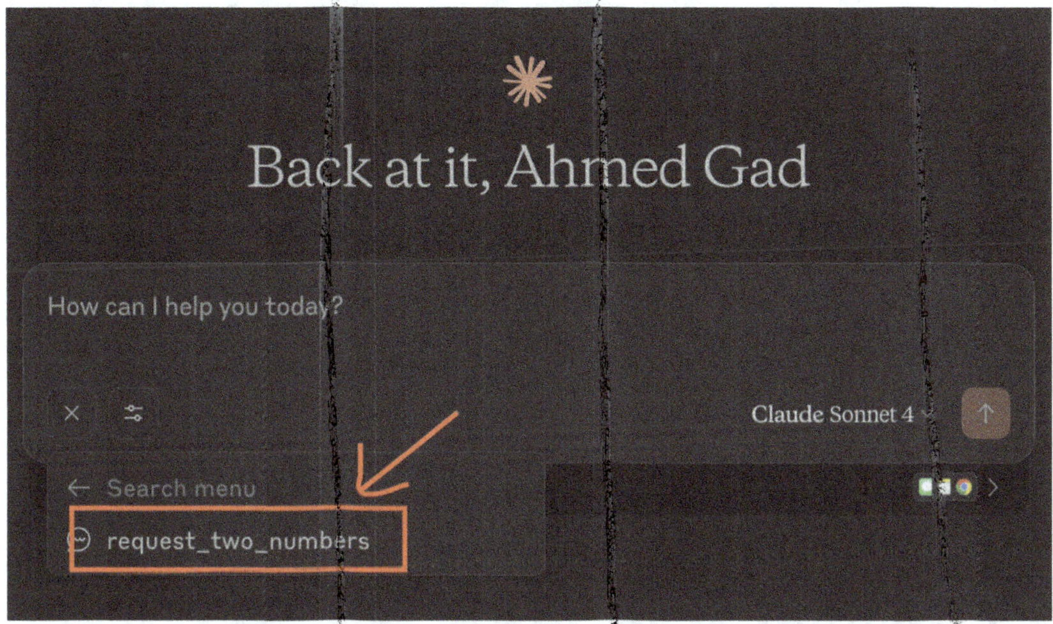

Figure 6-14. *The list of prompts inside the sum MCP server*

Since our MCP server prompt accepts arguments, then clicking it opens a new window where you can enter values for them as in Figure 6-15. Once these values are provided, they are sent to the MCP server, which generates the corresponding prompt for Claude to process.

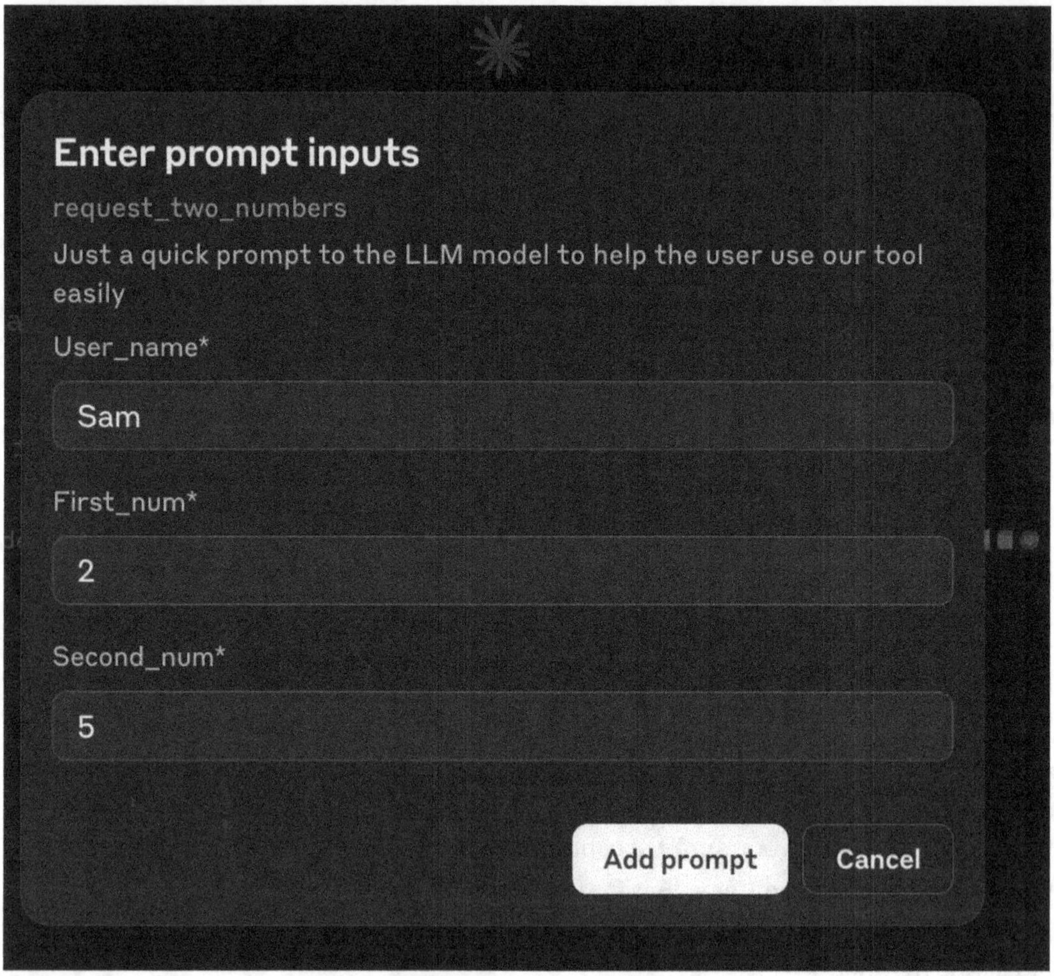

Figure 6-15. *Claude displays a window allowing the user to enter values for the arguments required by the selected prompt*

The prompt generated by the MCP server is returned to Claude, ready for use as given below:

```
You are an assistant for the user Ahmed. Given the accepted two numbers 1.0
and 5.0, what is their summation? Please greet the user by its name.
```

The user simply needs to submit the prompt for processing. Claude analyzes the prompt, as usual, and determines that the add_two_numbers tool is required to handle it. The tool is then invoked, and the generated response is presented to the user based on the context provided, as shown in Figure 6-16.

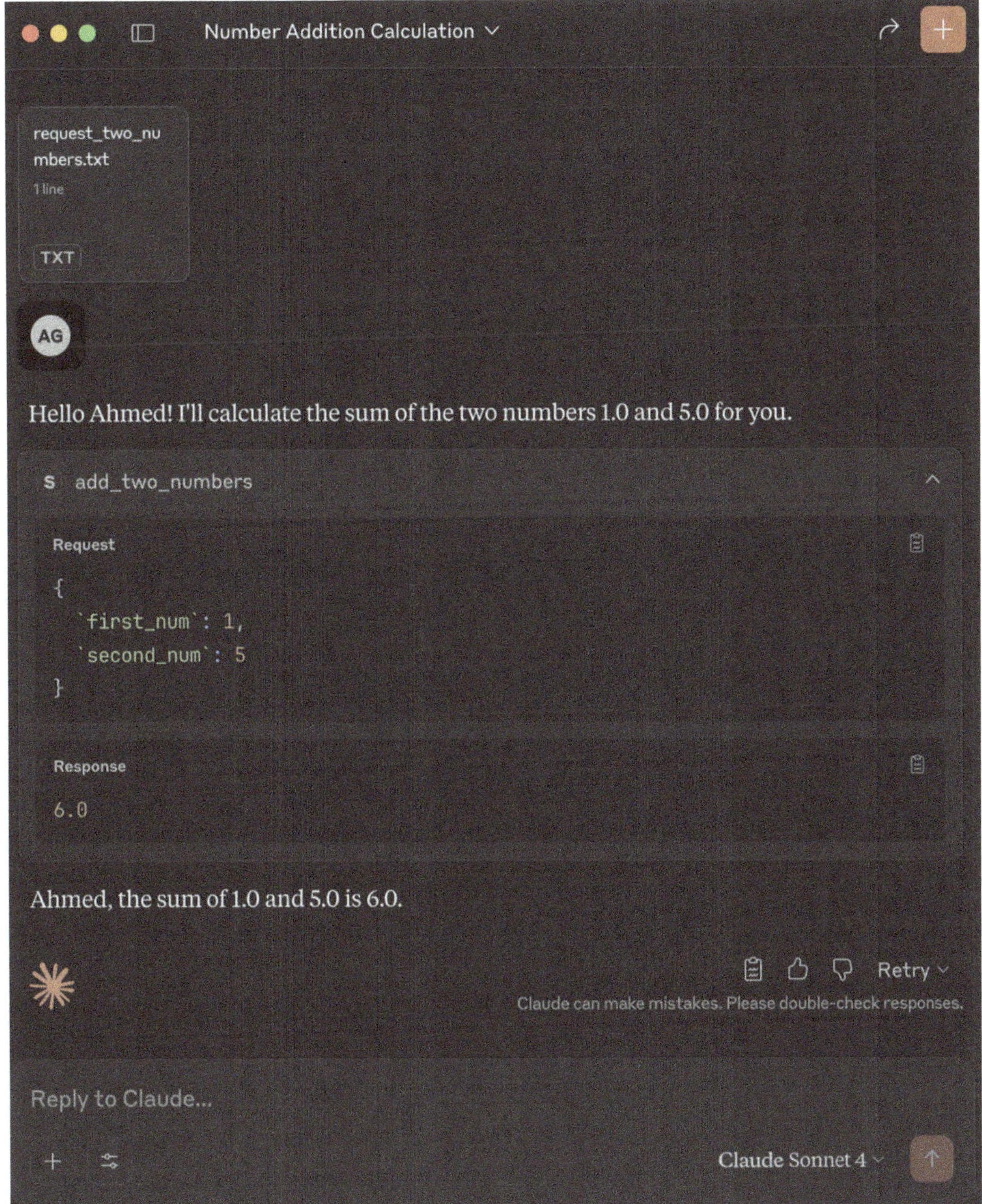

Figure 6-16. *Claude analyzes the prompt to call the appropriate MCP server tool*

This concludes our toy example. In the next section, we will build an MCP server designed to retrieve the latest football match results.

6.2.2 Football Match Result

To retrieve football match results, we will use the `API-Football` service available through RapidAPI. Access to this service requires an API key.

Follow these steps to obtain the API key:

1. Visit RapidAPI Hub `https://rapidapi.com/hub`.

2. Log in to your account or create a new one if you don't already have one.

3. Search for API-Football, or navigate directly to `https://rapidapi.com/api-sports/api/api-football`.

4. Click Subscribe to Test to begin using the API.

5. Select the free plan, which provides up to 100 requests, sufficient for testing the server.

6. After subscribing, go to the Endpoints tab.

7. Locate your API key under the name `X-RapidAPI-Key`.

With the API key obtained, you can proceed similarly to the previous steps, except for building a new MCP server tool that connects to the football data API and retrieves the match results.

Similar to the previous toy example, the first step in building the MCP server is to instantiate the `FastMCP` class. In this case, however, a new name and instruction are provided to reflect the specific purpose of the tool.

```python
from mcp.server.fastmcp import FastMCP

mcp = FastMCP(name="Football Results",
            instructions=("You are a football assistant that can fetch
            the latest match results for any club."))
```

Since HTTP requests will be made to an external server, the following helper function named `make_http_request()` is provided to handle those requests. Since it uses the `aiohttp` library, ensure that the uv project includes the `aiohttp` dependency. This is either by running the command `uv add aiohttp` or by manually adding it to the `pyproject.toml` file.

```python
import aiohttp

async def make_http_request(url: str, headers: dict) -> dict:
    async with aiohttp.ClientSession() as session:
        async with session.get(url, headers=headers) as resp:
            if resp.status != 200:
                return None
            return await resp.json()
```

Here are some predefined variables to assist in constructing the HTTP requests. Make sure to assign your API key to the API_KEY variable.

```python
BASE_URL = "https://api-football-v1.p.rapidapi.com/v3"

# Replace with your API key
API_KEY = "YOUR-API-KEY-HERE"

headers = {
    "x-rapidapi-host": "api-football-v1.p.rapidapi.com",
    "x-rapidapi-key": API_KEY
}
```

The MCP server exposes a tool named get_last_match_result, whose implementation proceeds in two main steps, as in the next code:

1. First, it issues an HTTP request to look up the team's unique identifier from its name.

2. Then, using that identifier, it sends a second HTTP request to fetch the team's most recent match data.

Finally, the tool parses the returned JSON payload and delivers the latest score back to the MCP client for further processing.

```python
@mcp.tool()
async def get_last_match_result(team_name: str) -> str:
    """
    Get the latest football match result for a given team name.
```

```python
    Args:
        team_name: The name of the football team.
    """

    team_search_url = f"{BASE_URL}/teams?search={team_name}"
    team_data = await make_http_request(team_search_url, headers)

    if not team_data or not team_data.get("response"):
        return f"Could not find team named '{team_name}'."

    team_id = team_data["response"][0]["team"]["id"]

    fixtures_url = f"{BASE_URL}/fixtures?team={team_id}&last=1"
    fixture_data = await make_http_request(fixtures_url, headers)

    if not fixture_data or not fixture_data.get("response"):
        return f"No recent matches found for team '{team_name}'."

    match = fixture_data["response"][0]
    home = match["teams"]["home"]["name"]
    away = match["teams"]["away"]["name"]
    score_home = match["goals"]["home"]
    score_away = match["goals"]["away"]
    date = match["fixture"]["date"]

    result = f"""Date: {date}
Match: {home} vs {away}
Score: {score_home} - {score_away}
Winner: {match['teams']['home']['winner'] and home or away}
"""

    return result.strip()
```

After creating the function, the next step is to run the MCP server, update the `claude_desktop_config.json` file with the new server details, and restart Claude. These steps ensure that Claude is configured to use the new tool.

Since no predefined prompt was created for this server, the user must manually enter a prompt relevant to the tool. For example:

```
What was the last match result for Manchester United?
```

As shown in Figure 6-17, Claude correctly identified that the tool `get_last_match_result` was required to respond to this query. It automatically issued a request using the team's name, *Manchester United*, to the MCP server tool. The tool's response provided Claude with sufficient context to generate a complete answer, stating that the team's most recent match ended in a 2–2 draw.

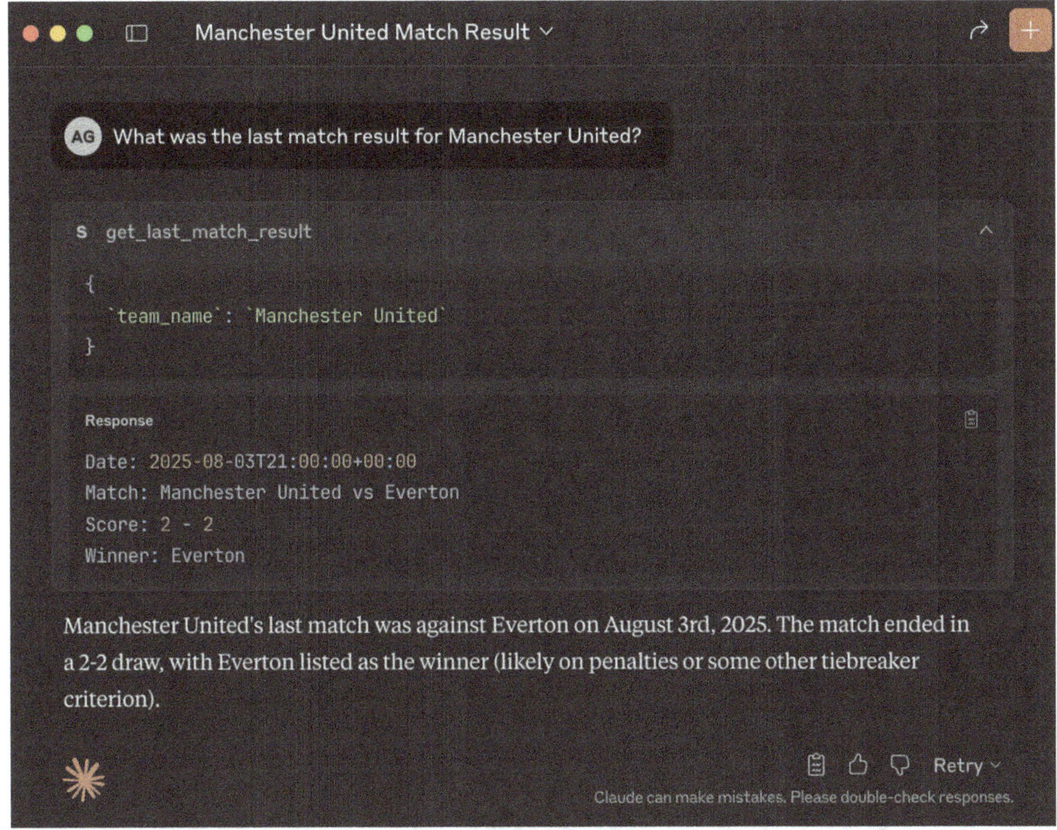

Figure 6-17. *Claude identifies the* `get_last_match_result` *MCP server tool as essential for retrieving the necessary context to answer a prompt about Manchester United's latest football match result*

This marks the end of the football match result example. Before moving on, it's worth highlighting that MCP servers offer several advanced features that can enhance the development of MCP-based applications. Two such features are elicitation and sampling, which expand the interaction dynamics between the MCP client and server.

6.2.3 Elicitation

Elicitation enables the MCP server to actively request additional input from the user via the MCP client. This is particularly useful in scenarios where a simple request-response cycle is insufficient. For instance, consider a tool that facilitates online transactions. Before completing a payment, the MCP server might issue an elicitation request asking the user to confirm the purchase with a "yes" or "no" response.

6.2.4 Sampling

Another powerful feature is sampling, which allows the MCP server to initiate an action by requesting a response from an LLM through the MCP client. This approach reverses the typical flow, where the client initiates requests, and the server merely responds. In the case of sampling, the MCP client acts as both a client and a server.

Sampling is especially valuable for delegating complex tasks (e.g., data analysis) to an LLM. The LLM generates a response based on the server's request, but since the server cannot dictate the exact model to be used, it is up to the client to choose the most suitable model available in its environment.

It's important to note that not all MCP clients support elicitation or sampling. A comprehensive list of supported features across various MCP clients can be found at modelcontextprotocol.io/clients. For example, Anthropic Claude does not currently support either elicitation or sampling, whereas VS Code GitHub Copilot supports both features and offers additional capabilities.

6.3 Agent2Agent (A2A) Protocol

The MCP plays a crucial role in enabling AI agents to access external resources and tools, enriching their context with relevant information to support more informed, knowledge-centric responses. However, MCP's usage is limited to assisting an agent in generating a response or taking a specific action. It does not inherently support collaboration or knowledge sharing between agents. Relying solely on MCP assumes that the agent either possesses sufficient knowledge to address any user prompt or that the user is already aware of the agent's capabilities when formulating their requests.

In reality, the AI agent might not have the capability to answer certain questions. Such questions or queries cannot be answered simply by seeking help from an external tool because the agent itself is not aware of the topic and does not know what actions it should take to handle the query. The use of MCP tools is limited only to fill out some specific gaps, not to assist in handling an entire query or question.

Think of it as a customer service agent who receives a question about a highly specialized product they've never heard of. Even if they have access to a database, they may not know what to search for or how to interpret the information if they don't understand the domain.

For example, imagine a user asking a general-purpose AI assistant:

What's the difference between homomorphic encryption and secure multi-party computation?

If the assistant weren't trained on advanced cryptography topics and didn't recognize either term, it wouldn't know how to phrase a query to an external tool or even realize that this is a cryptography-related question. In such cases, MCP alone falls short because it relies on the agent's existing knowledge and awareness to invoke the correct tools. But if the agent is aware of these topics and only has a missing piece of information such as the current performance benchmarks of a specific cryptographic method, then it can formulate a query to an MCP tool to address it.

This limitation highlights the need for more dynamic coordination between agents, where one agent can defer to another with more specialized expertise. This is precisely the gap that the Agent2Agent (A2A) protocol aims to fill.

In April 2025, Google announced the A2A protocol in a blog post titled Announcing the Agent2Agent Protocol (A2A). The specification of the protocol is accessible at `https://a2a-protocol.org`. Similar to the motivation behind MCP to have a standard interface between the agents and the tools, A2A aims to have a standardized way of communication between different agents offered by different vendors. Each agent could be considered a blackbox.

To illustrate how the A2A protocol can be used to build real-world agentic AI systems, consider the following major complex task:

Book a nearby restaurant for tomorrow evening that serves delicious pizza.

Since no single agent might have enough knowledge or resources to handle it, it is better to be decomposed into smaller, manageable subtasks where each task is delegated to a specialized agent with domain-specific expertise.

This task involves multiple steps and requires different capabilities such as geolocation, search, natural language understanding, and scheduling. A single agent cannot effectively handle all these components on its own. Instead, a primary agent (also called a master agent) coordinates with other specialized agents using a protocol like A2A to complete the full task.

The interaction among agents unfolds in the following steps:

1. The primary agent in the agentic system receives the high-level request, which orchestrates the subtasks required to fulfill the goal.

2. The primary agent invokes the location agent to determine the user's current location using device data or IP-based geolocation. Once the location is obtained, it is passed to the next agent.

3. The primary agent sends the location to the restaurant search agent, requesting a list of nearby pizza restaurants. The response includes restaurants ranked by rating and number of reviews.

4. For each restaurant in the list, the primary agent calls the review analysis agent. This agent uses natural language processing to evaluate whether customers praise the pizza as *delicious*. If negative reviews mention poor pizza quality, that restaurant is excluded from further consideration.

5. Once a restaurant with positive reviews is found, the system presents it to the user for approval. If the user rejects it, the process is repeated for the next restaurant.

6. Upon user approval, the booking agent attempts to reserve a table for tomorrow evening. If no time is available, the system returns to the next restaurant candidate and repeats the process.

7. Once a restaurant with both a good reputation and an available time slot is found, the system sends a final confirmation to the user. The booking is completed, and the task is considered successfully fulfilled.

Such a multi-agent approach reflects how the agentic AI systems can be autonomous and collaborate to fulfill tasks.

In the A2A protocol, agents communicate using the JSON-RPC (Remote Procedure Call) API. This protocol allows an agent to send a request to another agent or server to invoke a specific function or method remotely. The data is structured in JSON format, and the requests are transmitted over HTTP or HTTPS.

In agentic AI systems, it is essential for each agent to have awareness of the other agents in the network. This awareness extends beyond simply knowing that the other agents exist, as it also includes a clear understanding of their capabilities. Such knowledge is critical for deciding when to involve another agent and how to interact with it effectively.

These questions are addressed through the use of an Agent Card, as illustrated in Figure 6-18. Each agent is associated with its own agent card, which is a structured description that serves as a profile for that agent. By maintaining accurate agent cards, a multi-agent system can make informed routing decisions by selecting the most suitable agent for a given subtask and ensuring that interactions between agents are efficient, consistent, and contextually appropriate.

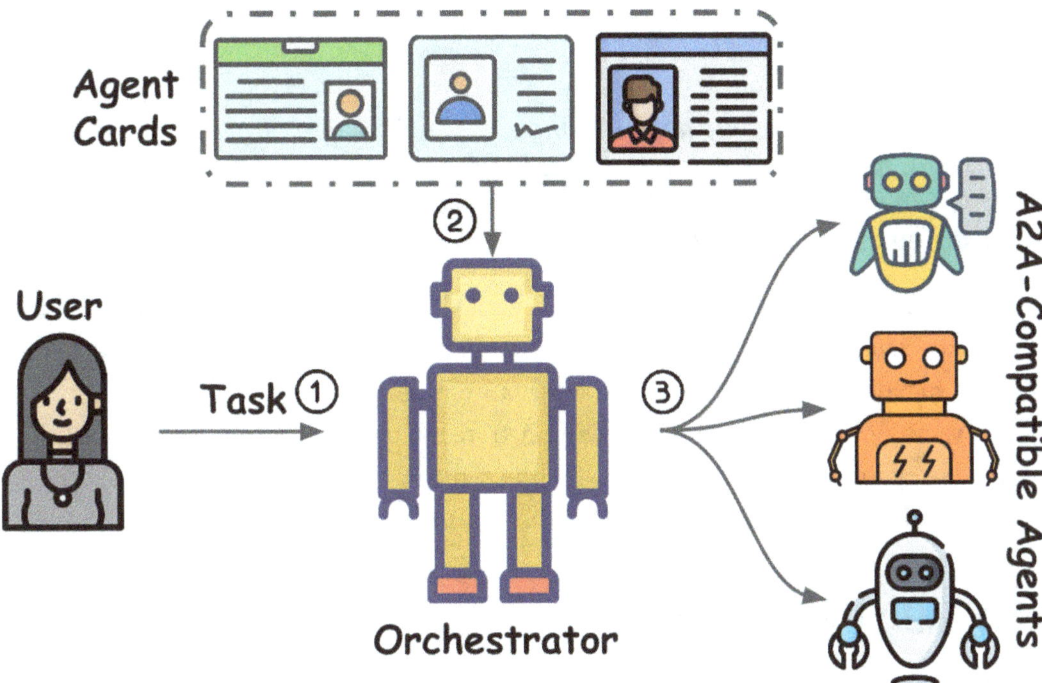

Figure 6-18. *A2A uses an orchestrator agent, which has the list of agent cards defining the other A2A-compatible agents' capabilities. Based on the request coming from the user, it decides the best agent to forward the request*

An agent card typically contains:

- **Name**: The agent's unique identifier.

- **Skills/Capabilities**: The domains or tasks the agent can handle.

- **Example Prompts**: Sample inputs the agent can process effectively.

- **Access Link**: The address where the agent can be reached.

- **Input/Output Data Format**: The expected structure of data for requests and responses.

- **Additional Metadata**: Optional information such as version, description, and tags.

Once a user submits a task, the orchestrator agent examines the available agent's cards to identify the agent whose skills are most relevant to the task. In some systems, this selection process is preceded or supplemented by a reasoning step in which the orchestrator evaluates whether the task can be decomposed into smaller, more manageable subtasks. For each subtask, it evaluates which agent should take responsibility.

The orchestrator agent does not maintain offline copies of Agent Cards. Instead, it stores only the network addresses of the other agents in the system. When needed, the orchestrator retrieves an agent's card on demand, ensuring that the most up-to-date profile is always used.

When an A2A client sends a request (i.e., a task) to an A2A-compatible agent using its address, the request is handled by an A2A server. The server forwards the request to the target agent, collects the response, and returns it to the client.

The remainder of this section will demonstrate the development of a simple A2A-compatible client–server application in Python. This example is intended to illustrate how the fundamental concepts of A2A can be applied in practice.

6.3.1 A2A Python SDK

The A2A protocol has software development kits (SDKs) for multiple programming languages, including Java, JavaScript, Python, and .NET. In this section, we focus on using the Python SDK to develop a simple project consisting of two independent agents:

1. **Weather Agent**: Responds to queries related to weather conditions.

2. **Recipe Agent**: Responds to queries related to food and cooking recipes.

Both agents follow the same development cycle. For each task submitted by the A2A client, we will examine how to select the most suitable agent based on the agent card of each agent.

In summary, the steps to create a client-server A2A-compatible application are:

1. **Define the Agent's Skills**: Specify the capabilities and jobs the agent can perform.

2. **Create the Agent Card**: Provide a structured description.

3. **Build the A2A Agent**: Implement the agent logic according to the defined skills.

4. **Implement the Agent Executor**: Create the component responsible for executing the agent's skills when invoked.

5. **Create a Request Handler**: It interacts with the agent's executor to handle the incoming A2A requests for the agent.

6. **Build the A2A Server**: Set up the server to host the agent and handle incoming requests.

7. **Run the Server**: Start the server so it can receive and process client requests.

8. **Develop the A2A Client**: Implement the client that will send tasks to the A2A server asking for some tasks to be accomplished by the A2A agent.

9. **Send Requests from the Client to the Server**: Test the full client–server interaction by submitting tasks and receiving responses.

Install the Libraries

The Python SDK for the A2A protocol is available in the official GitHub repository `https://github.com/a2aproject/a2a-python`. It can be installed directly with `pip`:

```
pip install a2a-sdk
```

For easier project management, we will use the *uv* tool, as discussed earlier in the context of MCP. Begin by initializing a new project (let's call it *a2aexample*):

```
uv init a2aexample
cd a2aexample
```

Next, create and activate a virtual environment:

```
uv venv
source .venv/bin/activate
```

Then, use the add command to install the required libraries. The a2a-sdk library provides the A2A functionality, while uvicorn is used to run the HTTP server. The openai library is utilized to interact with the API of OpenAI's GPT models. In this project, it is used to build a router component that determines which agent is best suited to handle each incoming task.

```
uv add "a2a-sdk"
uv add "uvicorn"
uv add "openai"
```

Throughout this section, whenever prompted to create a new Python script, place it in the root directory of the *uv* project or structure it the best way you like.

Agent Skill

Each A2A agent is defined by a set of skills representing specific tasks or capabilities it can perform. When an A2A client retrieves an agent's card, it can examine the listed skills to determine whether that agent is suitable for a given task.

A skill is created by instantiating the AgentSkill class, as in the next code. Save the code in a script file with a name of your choice, for example, a2a_server_weather.py.

The following properties describe the skill:

1. id: A unique identifier for the skill.

2. name: A human-readable name for the skill.

3. description: A detailed explanation of the skill, helping the A2A client assess the agent's scope and suitability for various tasks.

4. `tags`: Keywords used to organize and categorize skills. Tags can be shared among different agents to indicate overlapping or related capabilities.

5. `examples`: Sample prompts that can be used to test or demonstrate the skill.

6. `inputModes`: The supported data types for input, such as plain text, JSON, images, etc.

7. `outputModes`: The supported data types for output, similar to `inputModes` but describing the format of the agent's responses.

```python
from a2a.types import AgentSkill

skill = AgentSkill(id='weather',
                   name='Returns weather information',
                   description='Fetch the real-time weather prediction
                   information of a location',
                   tags=['weather', 'sunny', 'rainy', 'prediction'],
                   examples=['Is it sunny?', 'What is the weather today?'],
                   inputModes=['text/plain'],
                   outputModes=['text/plain'])
```

The values assigned to the skill clearly define its purpose, indicating that it is focused on weather prediction. In this project, each agent will be assigned a single skill.

Agent Card

Like how people use business cards for identification and to share contact information, A2A agents use agent cards to identify themselves, showcase their skills and capabilities, and provide their address to be reached by other agents.

An agent card is created using the `AgentCard` class as shown in the following code. Save it in the same `a2a_server_weather.py` script.

The arguments passed to the constructor of the `AgentCard` class include:

1. `name`: The agent's name.

2. `description`: A description of the agent to help other agents determine whether it is suitable for a task.

3. version: Since the agent may undergo changes or enhancements, the version argument helps select a specific version of the agent. This ensures that updates do not break other modules that depend on particular features or skills that might be removed in later versions.

4. url: The address of the agent's endpoint on the hosting server. If multiple agents run on the same machine sharing a domain, the port number distinguishes their URLs.

5. default_input_modes: The default media types accepted by the agent as input.

6. default_output_modes: Similar to default_input_modes, but for the agent's output formats.

7. capabilities: Indicates whether the A2A agent has special features, such as streaming support.

The agent's description is typically broad enough to encompass all the skills it supports, whereas each individual skill requires a more specific and detailed description.

```python
from a2a.types import AgentCapabilities, AgentCard

agent_port = 9999
public_agent_card = AgentCard(name='Weather Agent',
                              description='A weather agent',
                              version='1.0.0',
                              url=f'http://localhost:{agent_port}/',
                              default_input_modes=['text'],
                              default_output_modes=['text'],
                              capabilities=AgentCapabilities
                              (streaming=True),
                              skills=[skill])
```

The combination of a skill's name, description, tags, and examples, along with the agent's name, description, and version, provides sufficient information for any A2A client to determine whether the agent can perform a specific task. It is recommended to provide clear and precise values for these properties to avoid confusion.

Additionally, specifying the input and output modes for both the agent and its individual skills offers greater flexibility. While the agent itself may support multiple data types, particular skills can impose restrictions on the types they accept or produce.

Since the weather agent runs on port *9999* of the local host, its agent card is typically accessible at the URL:

http://localhost:9999/.well-known/agent-card.json

Note that this endpoint is accessible only when the server is running.

The agent card is represented as a JSON document, an example of which is provided below. Since no transport protocol was explicitly specified in the agent card and the A2A protocol communicates via the JSON-RPC API over HTTP, the `preferredTransport` field is automatically set to `JSONRPC`. All arguments passed to the `AgentSkill` class constructor are represented as JSON properties for each skill assigned to the agent.

```json
{
  "capabilities": {
    "streaming": true
  },
  "defaultInputModes": [
    "text"
  ],
  "defaultOutputModes": [
    "text"
  ],
  "description": "A weather agent",
  "name": "Weather Agent",
  "preferredTransport": "JSONRPC",
  "protocolVersion": "0.3.0",
  "skills": [
    {
      "description": "Fetch the real-time weather prediction information of
      a location",
      "examples": [
        "Is it sunny?",
        "What is the weather today?"
      ],
```

```
      "id": "weather",
      "inputModes": [
        "text/plain"
      ],
      "name": "Returns weather information",
      "outputModes": [
        "text/plain"
      ],
      "tags": [
        "weather",
        "sunny",
        "rainy",
        "prediction"
      ]
    }
  ],
  "url": "http://localhost:9999/",
  "version": "1.0.0"
}
```

Note that the agent card will be used later when we resume building the A2A server. For now, let's proceed with coding the agent and its executor before returning to the server's implementation.

Build the A2A Agent

The A2A agent is as simple as defining a class in Python and defining a method that will be called later when calling the agent. The next code gives a simple example of the weather agent as a class called *WeatherAgent*. Create a new Python file to save this code, for example, a2a_agent_executor_weather.py.

The behavior of the class is implemented in the invoke() method. This is a minimal agent implementation that does nothing except return a static text message. To incorporate an LLM or other AI capabilities, this method is where such functionality would be integrated.

```python
class WeatherAgent:
    """Weather Agent"""

    async def invoke(self):
        return 'I give weather prediction information'
```

Since agents could do some time-consuming tasks, their methods are typically defined as asynchronous (`async`). This includes querying external APIs or accessing databases. Using the `async` keyword in the method definition allows the agent to handle such operations without blocking the execution flow.

Agent Executor

The A2A agent itself is a simple Python class that does not rely on any A2A-specific features. The interaction with this agent is done via the agent executor. The next code gives a simple implementation of the agent executor.

A new class, `WeatherAgentExecutor`, extends the `AgentExecutor` interface and is created in the next code block. It is expected to be saved into the `a2a_agent_executor_weather.py` script.

The constructor creates an instance of the `WeatherAgent` class and assigns it to the agent instance attribute. This interface requires implementing two key methods that serve distinct purposes:

1. `execute()`: Receives incoming requests that expect responses from the agent. This method is responsible for invoking the agent to generate a response for the client.

2. `cancel()`: Handles cancellation requests for previously submitted tasks. This functionality is beyond the scope of this project.

```python
from a2a.server.agent_execution import AgentExecutor, RequestContext
from a2a.server.events import EventQueue
from a2a.utils import new_agent_text_message

class WeatherAgentExecutor(AgentExecutor):

    def __init__(self):
        self.agent = WeatherAgent()
```

```python
async def execute(self, context: RequestContext, event_queue:
EventQueue):
    result = await self.agent.invoke()
    await event_queue.enqueue_event(new_agent_text_message(result))

async def cancel(self, context: RequestContext, event_queue:
EventQueue):
    raise Exception('NA')
```

The execute() method accepts two parameters:

1. context: An instance of the RequestContext class containing information about the client's request, such as the original user prompt. We will explore the context's object content soon.

2. event_queue: An instance of the EventQueue class used for queuing the agent's responses to be sent back to the client.

Since the result of invoking the agent is a simple text message, it is wrapped using the new_agent_text_message() utility function. It returns a new Message object suitable for enqueueing onto the event queue and sending back to the client. This function takes a string as input and returns a new instance of the a2a.types.Message class. This is an example of the decoded result returned by the Message class.

The message includes a key called *parts* because a client's request can contain multiple parts with different data types, such as plain text, JSON, images, or other media. This design allows the system to handle complex requests composed of heterogeneous content. In this example, however, the request only includes a single part containing text data.

```json
{
  "contextId": null,
  "extensions": null,
  "kind": "message",
  "messageId": "0364cf02-6289-49a5-9058-7bd8253d157f",
  "metadata": null,
  "parts": [
    {
      "kind": "text",
      "metadata": null,
      "text": "I give weather prediction information"
```

```
    }
  ],
  "referenceTaskIds": null,
  "role": "agent",
  "taskId": null
}
```

The resulting message object is then passed to the event queue by calling the enqueue_event() method, which sends the response back to the A2A client.

This completes the a2a_agent_executor_weather.py script, which contains both the agent and its executor. The agent executor serves as a bridge between the user-defined A2A agent and the A2A server.

Request Handler

With the agent and executor classes implemented, the next step is to build the request handler. This will be an instance of the DefaultRequestHandler class, as shown in the following code example. The constructor of this class accepts two arguments: an agent executor object and a task store object. Save this code in the a2a_server_weather.py script.

```python
from a2a.server.request_handlers import DefaultRequestHandler
from a2a.server.tasks import InMemoryTaskStore
from a2a_agent_executor_weather import WeatherAgentExecutor

request_handler = DefaultRequestHandler(agent_executor=WeatherAgent
Executor(), task_store=InMemoryTaskStore())
```

When a task is received from an A2A client, it is passed to the request handler, which routes the JSON-RPC request to the appropriate method within the agent's executor. The task store is responsible for maintaining the status of tasks, which is particularly useful in stateful applications where tracking the progress of each task is essential.

A2A Server

With the agent card and request handler in place, the next step is to associate them using a server. To streamline this process, the a2a-sdk library provides the A2AStarletteApplication class, which uses the Starlette web framework to quickly build an ASGI server.

```python
from a2a.server.apps import A2AStarletteApplication

server = A2AStarletteApplication(agent_card=public_agent_card,
                                 http_handler=request_handler)
```

Run the A2A Server

Once the server is defined, it can be run using the Uvicorn library. In this example, the server listens on port *9999*, which matches the port in the URL specified in the agent card. Using the IP address 0.0.0.0 means that requests arriving on port *9999* from any network interface of the local machine will be routed to the weather agent. The code should be saved in the a2a_server_weather.py script.

```python
import uvicorn

uvicorn.run(server.build(),
            host='0.0.0.0',
            port=agent_port)
```

You can run the server by using the uv tool:

```
uv run a2a_server_weather.py
```

At this point, we have built the following components:

1. Agent Skill

2. Agent Card

3. A2A Agent itself

4. Agent Executor

5. Request Handler

6. A2A Server

The interaction among these components is illustrated in Figure 6-19. First, the user submits a request to an A2A client (which will be implemented in the next section). The client maintains a list of A2A agent URLs. Using the URLs, it sends requests to retrieve the agents cards, which are essentially JSON files describing the agents. For example, the URL for the weather agent is: http://localhost:9999/.well-known/agent-card.json

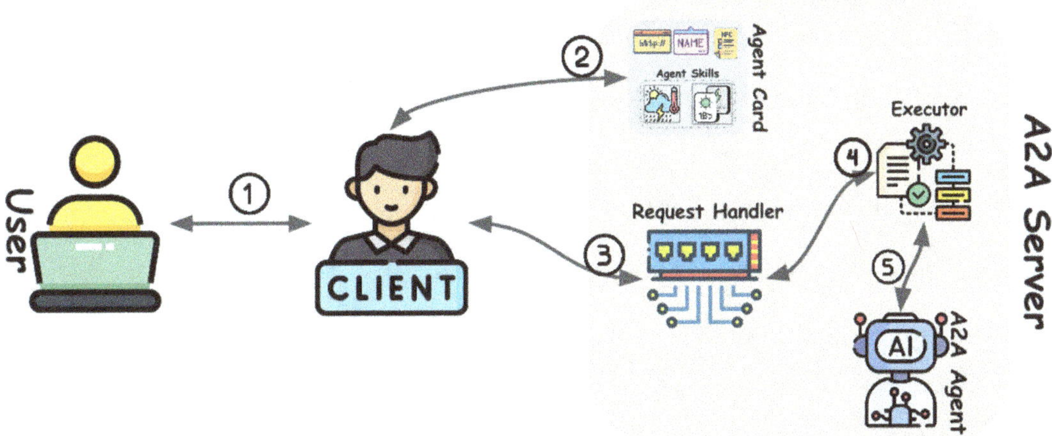

Figure 6-19. *Interaction among the various A2A components. The user sends a task to the client, which retrieves the agent card via an HTTP request. The client then submits the request to the A2A agent running inside the server, where it passes through the request handler and is processed by the agent executor*

Once the client receives the agents cards, it finds the best agent that handles the request. Assuming it selected the weather agent, then it sends a request to the A2A server to submit a task to the selected agent. The server's request handler receives the request and forwards it to the proper method inside the executor handler (e.g., execute()). The executor handler then invokes the agent to return a response. The agent's response is sent back to the client, which uses it to formulate a final response to the user.

Since the A2A server is now up and running, the final piece of the system is the A2A client.

A2A Client

The A2A client is responsible for submitting user requests, fetching the appropriate agent cards, and communicating with the A2A server to obtain responses. In this section, we will implement the A2A client and show how it interacts with the A2A server.

The following code initializes a list of agent URLs known to the client. Save this code in a new script, such as `a2a_client.py`.

When the client receives a request, it fetches the agent cards from these URLs to determine the most suitable agent for the task. Currently, there is only one agent on the list.

To send HTTP requests, the client uses the `AsyncClient` class from the `httpx` library, configured with a timeout of 10 seconds. This client is used within the `A2ACardResolver` class to fetch agent cards. Later, it will also be used to create the A2A client itself.

```python
from a2a.client import A2ACardResolver
import httpx

agents_base_urls = ['http://localhost:9999']

httpx_client = httpx.AsyncClient(timeout=10)

selected_agent_card = None
all_agents_cards = []

for agent_base_url in agents_base_urls:
    resolver = A2ACardResolver(httpx_client=httpx_client,
                               base_url=agent_base_url)
    try:
        agent_card = (await resolver.get_agent_card())
        all_agents_cards.append(agent_card)
    except Exception as e:
        raise RuntimeError('Failed to fetch the public agent card. Cannot
        continue.') from e
```

Once an agent card is retrieved, it is appended to the `all_agents_cards` list. Since it is possible that no agent card is fetched, it is important to check whether the list contains at least one agent card before proceeding.

```python
if len(all_agents_cards) == 0:
  raise RuntimeError('Failed to find any agent card.')
```

For now, we will hardcode the selected agent to always be the first agent in the list. In a later step, we will enhance this selection process by using one of OpenAI's GPT models to choose the most appropriate agent dynamically.

```python
selected_agent_card = all_agents_cards[0]
```

Once an agent card is retrieved, the next step is to create an instance of the
A2AClient class to establish a connection with the corresponding A2A agent.

```python
from a2a.client import A2AClient

client = A2AClient(httpx_client=httpx_client,
                   agent_card=selected_agent_card)
```

Send Requests

The A2A client now has an active connection to the A2A server and is ready to send
requests. The following code prepares a message payload as a JSON object containing a
prompt asking about the weather.

In this payload, the role is set to "user", indicating that the message is sent by the
user. The message includes only a single part of the kind "text", which contains the
prompt. This aligns with the expectations defined in the agent's agent card and skill,
both of which handle only text data types.

*The role field is set to agent in the generated response, indicating that the
message originates from the A2A agent.*

Finally, a unique messageId is assigned to the message to enable tracking or
referencing the message if needed.

```python
from uuid import uuid4

prompt = 'What is the weather in Cairo?'

send_message_payload = {
    'message': {
        'role': 'user',
        'parts': [{'kind': 'text', 'text': prompt}],
        'messageId': uuid4().hex,
    },
}
```

To communicate with the A2A server, the message must be formatted in the
expected data structure. This is done by creating an instance of the SendMessageRequest
class. Each request is assigned a unique ID, which can be used to track the status of the
request or to cancel it if needed.

The message JSON object is first wrapped into a `MessageSendParams` instance, which encapsulates the message details in the required format. Then, the client sends the request by calling the `send_message()` method, passing the `SendMessageRequest` object. This method handles the transmission of the message to the A2A server. The client has to wait for the agent to return a response.

```python
from a2a.types import MessageSendParams, SendMessageRequest

request = SendMessageRequest(id=str(uuid4()),
                             params=MessageSendParams(**send_message_
                             payload))

response = await client.send_message(request)
print(response.model_dump(mode='json', exclude_none=True))
```

This is an example of the request returned from the `SendMessageRequest` class.

```json
{
  "id": "dd683afe-819e-4a0b-8be6-90da3156b9b3",
  "jsonrpc": "2.0",
  "method": "message/send",
  "params": {
    "message": {
      "kind": "message",
      "messageId": "d79d1008c89241059c9c98ad79db9fb0",
      "parts": [
        {
          "kind": "text",
          "text": "What is the weather in Cairo?"
        }
      ],
      "role": "user"
    }
  }
}
```

The previous code should be placed inside a method, such as `main()`, to organize the asynchronous workflow properly.

```python
async def main():
    ...
```

Finally, the method is invoked asynchronously to send the request.

```python
if __name__ == '__main__':
    import asyncio
    asyncio.run(main())
```

The client is now ready, and we can run it using uv to interact with the server.

```
uv run a2a_client.py
```

When the client runs, it creates a request to the A2A server, which forwards it to the A2A agent. This is an example of the response received from the server.

```json
{
  "id": "dd683afe-819e-4a0b-8be6-90da3156b9b3",
  "jsonrpc": "2.0",
  "result": {
    "kind": "message",
    "messageId": "d1a7e502-60e7-4ad9-86e1-98dac45df1e3",
    "parts": [
      {
        "kind": "text",
        "text": "I give weather prediction information"
      }
    ],
    "role": "agent"
  }
}
```

Since the agent is currently hardcoded to always return the same response, the received message from the agent contains the text: "I give weather prediction information."

With the weather agent complete, the basic A2A workflow is now clear, from running an agent inside a server to creating a client that invokes the agent. We created the following Python scripts for building the client and server components:

1. `a2a_server_weather.py`: Implements the A2A agent's skill, agent card, and server.

2. `a2a_agent_executor_weather.py`: Defines the A2A agent and its executor logic.

3. `a2a_client.py`: Implements the A2A client responsible for sending requests and receiving responses.

Now, let's turn our attention to building the recipe agent. Unlike the weather agent, this agent uses OpenAI's GPT model to generate dynamic, context-aware responses based on user prompts.

6.3.2 Recipe Agent

The second recipe agent will almost follow the same steps we did for the weather agent except for some changes to make it use one of the OpenAI GPT models. Based on the previous workflow in Figure 6-19, the new workflow is summarized in Figure 6-20 to highlight the core changes before diving into the development details.

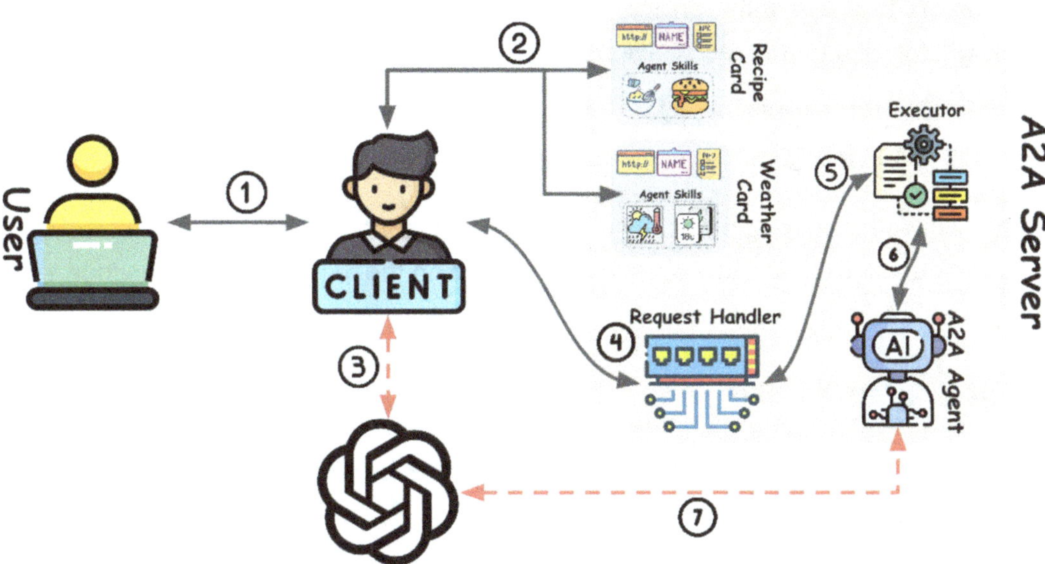

Figure 6-20. *Interaction among the various A2A components after using GPT to select the agent. The user sends a task to the client, which retrieves the agent cards via an HTTP request. GPT evaluates these cards to identify the most suitable agent. The client then submits the request to the GPT-selected A2A agent. On the server side, the request is received by the request handler and processed by the agent executor, which passes it to the agent that also uses GPT to generate a response to the user's query*

There are two key modifications:

1. To determine which agent the client should use, the client sends a prompt to GPT, asking it to evaluate the available agents based on their cards. GPT responds with the index of the most appropriate agent.

2. When the selected A2A agent receives the request through the agent's executor, it also uses GPT to generate the final response.

Instead of going through all the steps again, we will only focus on the areas where the changes are necessary.

Compared to the Python script names used for the weather agent, the following names will be used for the recipe agent:

1. a2a_server_recipe.py

2. a2a_agent_executor_recipe.py

We will continue using the same a2a_client.py script, editing it to enable routing the user's questions to the appropriate agent based on the request. The major code changes will be in the client.

Agent Skill and Card

Since we are tackling a new agent with new responsibilities, then the first change is definitely in the agent skills and card. The next code defines the single skill and card of the recipe agent. Similar to the weather agent, it also accepts and returns plain text. But the agent is now running on port *8888*.

```python
from a2a.types import AgentCapabilities, AgentCard, AgentSkill

skill = AgentSkill(id='recipe',
                   name='Returns recipe information',
                   description='just returns recipe information',
                   tags=['recipe', 'food', 'lunch'],
                   examples=['How to make a pizza?', 'Suggest a recipe'],
                   inputModes=['text/plain'],
                   outputModes=['text/plain'])

agent_port = 8888
public_agent_card = AgentCard(name='Recipe Agent',
                              description='A recipe agent',
                              url=f'http://localhost:{agent_port}/',
                              version='1.0.0',
                              default_input_modes=['text'],
                              default_output_modes=['text'],
                              capabilities=AgentCapabilities
                              (streaming=True),
                              skills=[skill])
```

Agent and Executor

Each agent in an agentic AI system specializes in its own skill set. This specialization can arise from several factors, such as:

- Using an LLM that has been specifically trained or fine-tuned to perform a particular task with high efficiency and quality.

- Employing a general-purpose LLM but augmenting it with external tools that provide comprehensive context for answering questions.

- Even using a standard LLM guided by carefully crafted prompts that make the model function in a specific way.

In this demo, the recipe agent simply uses a plain GPT model without any additional fine-tuning or enhancements. In real-world applications, agents typically incorporate customizations to excel at their designated tasks with cutting-edge performance.

For the recipe agent, we create a new class called `RecipeAgent`, which utilizes the OpenAI API to call the GPT-3.5 Turbo model. This agent generates responses based on the prompts it receives. To access this model, you need your own OpenAI API key. However, as discussed in earlier chapters, you can also use free models available on platforms like Hugging Face.

```python
from a2a.server.agent_execution import AgentExecutor, RequestContext
from a2a.server.events import EventQueue
from a2a.utils import new_agent_text_message

from openai import OpenAI
client = OpenAI(api_key="sk-...")

class RecipeAgent:
    """Recipe Agent."""

    async def invoke(self, prompt) -> str:
        response = client.responses.create(model="gpt-3.5-turbo",
                                           input=prompt)
        return response.output_text
```

The agent executor for the recipe agent is identical to that of the weather agent, with the only difference being that the `agent` instance attribute is set to an instance of the newly created `RecipeAgent` class.

```python
class RecipeAgentExecutor(AgentExecutor):

    def __init__(self):
        self.agent = RecipeAgent()

    async def execute(self, context: RequestContext, event_queue:
    EventQueue) -> None:
        result = await self.agent.invoke(prompt=context.get_user_input())
        await event_queue.enqueue_event(new_agent_text_message(result))

    async def cancel(self, context: RequestContext, event_queue:
    EventQueue) -> None:
        raise Exception('cancel not supported')
```

Server

The third change is in the server, where it replaces the previous executor with the new RecipeAgentExecutor class. Aside from this update, everything else remains identical.

```python
from a2a.server.apps import A2AStarletteApplication
from a2a.server.request_handlers import DefaultRequestHandler
from a2a.server.tasks import InMemoryTaskStore
from a2a_agent_executor_recipe import RecipeAgentExecutor
import uvicorn

request_handler = DefaultRequestHandler(agent_executor=RecipeAgent
Executor(), task_store=InMemoryTaskStore())

server = A2AStarletteApplication(agent_card=public_agent_card,
                                 http_handler=request_handler)

uvicorn.run(server.build(),
            host='0.0.0.0',
            port=agent_port)
```

Client

The client component will have significant changes and additions. One key addition is a new function named select_best_agent(), which accepts two arguments:

1. prompt: The user's input prompt.

2. agents_cards: A list containing all agent cards known to the client.

This function uses the advanced GPT-4.1 model to determine which agent is best suited to handle the given prompt. The prompt is carefully engineered so that the LLM returns a single number corresponding to the index of the most appropriate agent in the list. If no suitable agent is found, the function returns -1.

```python
def select_best_agent(prompt, agents_cards):
    card_selection_prompt = f"""
Given the following user prompt: "{prompt}", choose the best matching
agent card from the list below:

    {agents_cards}

Just return the index of the best matching agent. Return -1 if no agent
is suitable.
"""

    response = client.responses.create(model="gpt-4.1",
                                       input=card_selection_prompt)

    try:
        agent_id = int(response.output_text)
        if 0 <= agent_id <= len(agents_cards) - 1:
            pass
        elif agent_id == -1:
            raise Exception("No agent is capable to respond to the
            prompt.")
        else:
            raise Exception("Unexpected agent ID.")
```

```
except:
    raise Exception("Unexpected response from the LLM")

return agent_id
```

Based on the response from the model, the function validates the agent index. If the index is valid, it returns it; otherwise, it raises an exception indicating an invalid selection. Now, let's proceed to edit the main() method to incorporate agent card resolution and the selection of the best agent based on the user's prompt.

The updated main() method begins by resolving the agent cards from the provided agent URLs. If no agent cards are found, it raises an exception. Otherwise, it calls the select_best_agent() function with the user's prompt and the list of agent cards to identify the most suitable agent to handle the request.

```
async def main():
    agents_base_urls = ['http://localhost:9999',
                        'http://localhost:8888'
                        ]

    httpx_client = httpx.AsyncClient(timeout=10)


    selected_agent_card = None
    all_agents_cards = []

    for agent_base_url in agents_base_urls:
        resolver = A2ACardResolver(httpx_client=httpx_client,
                                   base_url=agent_base_url)

        try:
            agent_card = (await resolver.get_agent_card())
            all_agents_cards.append(agent_card)
        except Exception as e:
            raise RuntimeError('Failed to fetch the public agent card.
            Cannot continue.') from e

    if len(all_agents_cards) == 0:
        raise RuntimeError('Failed to find any agent card.')
```

```
prompt = 'Suggest a good pasta recipe'
selected_agent_index = select_best_agent(prompt=prompt,
                                    agents_cards=all_agents_cards)
...
```

The selected agent card's index is used to retrieve the corresponding card from the list. This card is then used to create an instance of the `A2AClient` class, establishing a connection to the chosen agent.

Next, the message is prepared as a JSON object, and a request is constructed using the `SendMessageRequest` class. The client then sends this request to the agent for processing using the `send_message()` method.

```
async def main():
    ....

    selected_agent_card = all_agents_cards[selected_agent_index]

    client = A2AClient(httpx_client=httpx_client,
                    agent_card=selected_agent_card)

    send_message_payload = {
        'message': {
            'role': 'user',
            'parts': [{'kind': 'text', 'text': prompt}],
            'messageId': uuid4().hex,
        },
    }

    request = SendMessageRequest(id=str(uuid4()),
                                params=MessageSendParams(**send_message_
                                payload))

    response = await client.send_message(request)
```

Finally, the `main()` function is called to start the agent's execution.

```
if __name__ == '__main__':
    import asyncio
    asyncio.run(main())
```

Given that the current prompt, *"Suggest a good pasta recipe,"* relates to the recipe agent's domain, GPT selected this agent to handle the user's request. This is the response received from the A2A agent:

```
{
  "id": "b090af52-aa95-4bfa-ad82-3713eb51e2d5",
  "jsonrpc": "2.0",
  "result": {
    "kind": "message",
    "messageId": "c409d69b-a45e-451c-a9bd-3c27a3e3d76d",
    "parts": [
      {
        "kind": "text",
        "text": "Sure! How about trying a delicious and flavorful recipe
        for Garlic Butter Shrimp Pasta? Here's how you can make it:\n\
        nIngredients:\n- 8 oz pasta of your choice (such as spaghetti or
        fettuccine)\n- 1 lb large shrimp, peeled and deveined\n- 4 cloves
        of garlic, minced\n- 4 tbsp unsalted butter\n- 1/2 cup chicken
        broth\n- 1/4 cup heavy cream\n- 1/4 cup grated Parmesan cheese\n-
        Salt and pepper to taste\n- Red pepper flakes (optional)\n- Fresh
        parsley, chopped for garnish\n\nInstructions:\n1. Cook the pasta
        according to package instructions until al dente. Drain and set
        aside.\n2. In a large skillet, melt the butter over medium heat.
        Add the minced garlic and sauté for about 1 minute until fragrant.\
        n3. Add the shrimp to the skillet and cook for 2-3 minutes on
        each side until they turn pink and opaque. Remove the shrimp from
        the skillet and set aside.\n4. Pour the chicken broth into the
        skillet and let it simmer for a few minutes. Stir in the heavy
        cream and Parmesan cheese, and season with salt, pepper, and red
        pepper flakes if using.\n5. Add the cooked pasta and shrimp back
        into the skillet and toss everything together until well coated
        in the sauce.\n6. Cook for another minute or two until everything
        is heated through.\n7. Serve the Garlic Butter Shrimp Pasta hot,
        garnished with chopped parsley.\n\nEnjoy your delicious Garlic
        Butter Shrimp Pasta!"
      }
```

```
    ],
    "role": "agent"
  }
}
```

6.4 Google Agent Development Kit

Having explored how to connect applications using the model context protocol (MCP) and Agent-to-Agent (A2A) protocols, we now turn our attention to the brain of the operation, which is the agent. While protocols standardize how data is exchanged, the agent remains the central entity responsible for reasoning and decision-making.

To streamline this process, Google introduced the Agent Development Kit (ADK), a high-level framework designed to build, manage, and deploy AI agents. The ADK is model agnostic, though it offers seamless integration with Google's Gemini models. The ADK is well-documented at this page `https://google.github.io/adk-docs`.

The ADK provides a dedicated Python library called `google-adk`. In the following examples, we will use version 1.27.2. To install it, run:

```
pip3 install google-adk
```

6.4.1 Build an Agent Using Google ADK

Once installed, then create a directory with the following files, which are used by the library:

1. `agent.py`: The core logic where the agent and its tools are defined.

2. `.env`: Stores configuration environment variables.

3. `__init__.py`: Marks the directory as a Python package.

You can quickly create these files using this command:

```
adk create test_agent
```

You will be prompted to select some properties about the agent. The first question is about which model to use by the agent. Since it is developed by Google, it has easy integration with Google models. But the library is model agnostic and still supports models by other providers.

Let's choose the first option to use the `gemini-2.5-flash` model for experimentation.

```
Choose a model for the root agent:
1. gemini-2.5-flash
2. Other models (fill later)
Choose model (1, 2): 1
```

Then you will be asked about the backend to use. To quickly get started, let's use Google AI.

```
1. Google AI
2. Vertex AI
Choose a backend (1, 2): 1
```

By selecting Google AI as the backend, you will be asked for the Gemini API key. If you do not have a key, just visit this page in Google AI Studio to create a key: `https://aistudio.google.com/app/apikey`

```
Enter Google API key: ****
```

Now, the 3 files listed above will be created. The `.env` file has the API key set to the `GOOGLE_API_KEY` environment variable. The `__init__.py` script will have a single line to import the agent script.

```python
from . import agent
```

The `agent.py` script is where the agent's actual code exists. This is the default code in the script. As expected, the model property of the model is set to `gemini-2.5-flash` as instructed previously. The `instruction` is the system instruction passed alongside the user's prompt to the LLM model. It must be descriptive enough to reflect the agent's scope.

```python
from google.adk.agents.llm_agent import Agent

root_agent = Agent(
    model='gemini-2.5-flash',
    name='root_agent',
```

```
    description='A helpful assistant for user questions.',
    instruction='Answer user questions to the best of your knowledge',
)
```

This agent does not yet have any tools. Using the `tools` property, we can assign the agent the actions and tools it can use to handle the prompts. This can be either a Python function or an instance of a class inheriting the `google.adk.tools.base_toolset.BaseTool` class, or an instance of another agent.

Let's use the simplest option to build a Python function that does nothing except for returning Hello.

```python
from google.adk.agents.llm_agent import Agent

def say_hello():
  """Greeting the user by saying Hello."""
  return "Hello"

root_agent = Agent(
    model='gemini-2.5-flash',
    name='root_agent',
    description='An agent that greets users.',
    instruction='You are an agent that greets the user. Whenever you
    receive a welcoming prompt, greet the user.',
    tools=[say_hello]
)
```

6.4.2 Run Agent

To interact with your agent via the command line, use the `adk run` command from the parent directory:

```
adk run test_agent
```

The user will be prompted to enter a message. When the message is passed, there is a flow that must be followed. In summary, the Runner acts as the Google ADK's engine and the primary interface, which captures the raw input and passes it to a root agent.

The root agent decides if it needs to execute other tools. Given the description assigned to our agent, it helps to decide the best tool to invoke. The request is passed to the proper tool, which returns a response that is finally served back to the user.

This is how our agent responded when the user entered the message *Hey there*. Since the user passed a relevant prompt, the agent responded with the static response of *Hello*.

```
Running agent root_agent, type exit to exit.
[user]: Hey there
[root_agent]: Hello
```

However, if a prompt falls outside the agent's defined scope, the agent will correctly decline the request:

```
[user]: What is the sum of 2 and 5?
[root_agent]: I am sorry, I cannot answer that question. I can only
greet users.
```

Beyond the CLI, the ADK includes a built-in web server. This allows you to test your agent in a browser-based chat interface. To launch the web application on port 8000 using a Uvicorn-based server, run:

```
adk web --port 8000
```

You can then access the agent at `http://127.0.0.1:8000` as shown in Figure 6-21. The underlying logic remains identical. The orchestrator receives the prompt, selects the tool, and returns the response through the web UI.

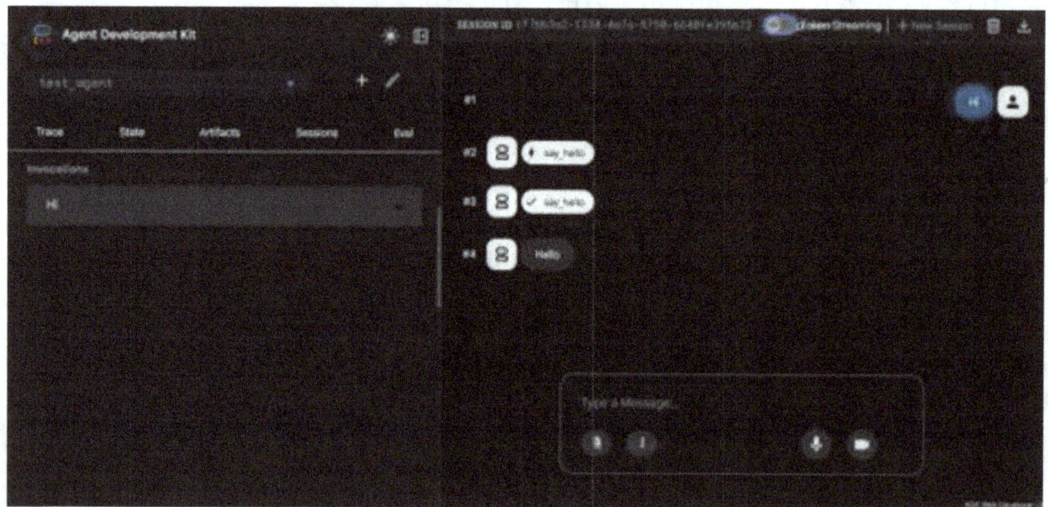

Figure 6-21. *Access a Google ADK agent through the web interface*

6.4.3 Tools with Arguments

So far, the agent has only a single tool that does not receive any arguments. This section extends the agent to build another agent that sums two numbers.

According to the next code, a new function called sum_nums() accepts two arguments representing the numbers to sum and returns the sum result. It is documented to help the agent figure out its capabilities. The description and the instruction properties of the agent are adjusted to reflect the two tools it supports.

```python
from google.adk.agents.llm_agent import Agent


def say_hello():
  """Greeting the user by saying Hello."""
  return "Hello"

def sum_nums(num1, num2):
    """Sum two numbers.

    Args:
        num1: The first number.
        num2: The second number.

    Returns:
        Sum of the two numbers.
    """

    return num1 + num2

root_agent = Agent(
    model='gemini-2.5-flash',
    name='root_agent',
    description='An agent that greets users and sums two numbers.',
    instruction='You are an agent that greets the user and sums two
    numberss.',
    tools=[say_hello, sum_nums]
)
```

When prompted, the prompt is passed to the sum_nums tool. The root agent uses our agent's result to respond to the user.

```
Running agent root_agent, type exit to exit.
[user]: What is 2+3?
[root_agent]: The sum of 2 and 3 is 5.
```

While these examples cover the fundamentals of the Google ADK, they represent only the entry point into a much deeper ecosystem of agentic workflows. By extending the capabilities of the root_agent, you can transform a simple greeter into a sophisticated autonomous assistant through methods such as accessing external tools using APIs or building multiple agents.

Index

F

G

S

T

GPSR Compliance
The European Union's (EU) General Product Safety Regulation (GPSR) is a set
of rules that requires consumer products to be safe and our obligations to
ensure this.

If you have any concerns about our products, you can contact us on

ProductSafety@springernature.com

In case Publisher is established outside the EU, the EU authorized
representative is:

Springer Nature Customer Service Center GmbH
Europaplatz 3
69115 Heidelberg, Germany